John H. Tullock

Water Chemistry
for the Marine Aquarium

Everything About Seawater, Cycles, Conditions, Components, and Analysis

Filled with Full-color Photographs

Illustrations by Michele Earle-Bridges

BARRON'S

CONTENTS

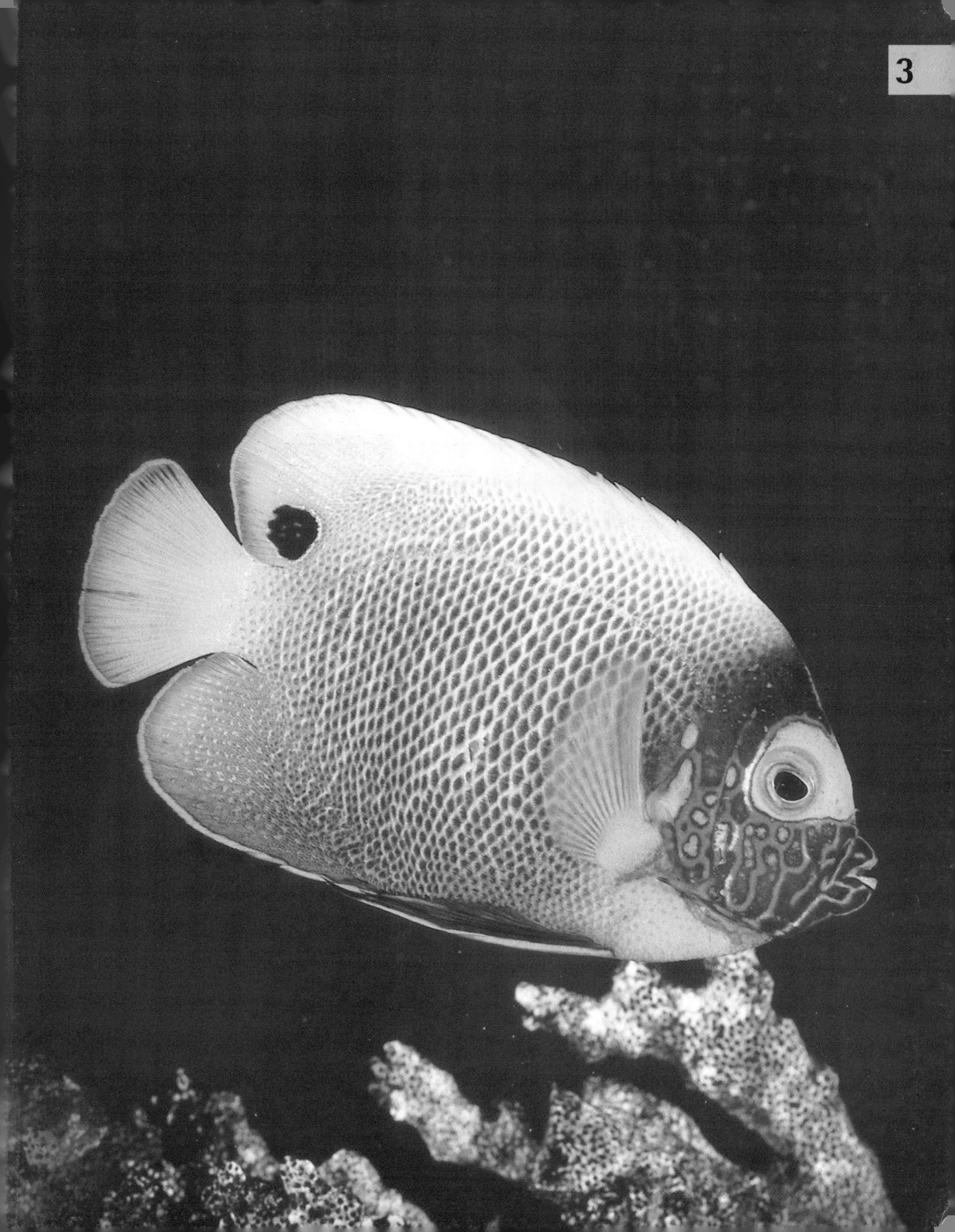

Coral reefs are the source of colorful fish and invertebrates for marine aquariums. Coral reefs are found only where specific physical and chemical conditions occur in the sea. Understanding the reef environment is fundamental to successfully maintaining coral reef organisms in captivity.

The Marine Environment

A widely held perception exists among aquarium hobbyists that marine fish are less hardy than freshwater fish. However, the difficulty associated with maintaining appropriate water conditions in the seawater aquarium probably causes more specimen losses than any inherent nonadaptability of marine fish. Although the differences between marine and freshwater environments are significant, keeping a marine aquarium should actually be simple because the focus is upon a single habitat, the coral reef. Freshwater aquariums, by contrast, may house fish or plants from a variety of habitat types. These may be quite different from each other with respect to the physical and chemical parameters of the water. Knowledge of basic seawater chemistry leads the aquarist to a better understanding of marine

Coral reef fishes and invertebrates are the "stars" of the marine aquarium.

habitats and provides the most reliable guide to marine aquarium maintenance.

Understanding what makes the marine environment unique also provides insight regarding what makes marine fish and invertebrates so different from their freshwater counterparts. Freshwater habitats are characterized by significant variations in pH, dissolved substances, and temperature, depending upon where on the planet they happen to occur. Even though most freshwater aquariums are restricted to tropical species, conditions in an Amazonian rain forest stream are entirely different from those in Africa's Lake Tanganyika, although both are freshwater habitats. Aside from temperature, seawater, by contrast, is remarkably similar in terms of its chemical and physical properties, whether bathing a coral reef or the Aleutian Islands.

Important physiological differences exist between marine fish and freshwater species. Both types have body fluids that are similar in composition. In freshwater species, the surrounding water, although containing some dissolved substances, is less salty than the body fluids. The opposite situation exists for marine fish. When two solutions of differing salt concentrations are separated by a semipermeable membrane, such as the skin of a fish, the water tends to pass across the membrane from the area of lower salt concentration to the area of higher salt concentration until the concentration on either side is equalized. This tendency is known as *osmotic pressure*, and the movement of water across the membrane is *osmosis*. Freshwater fish are said to be *hyperosmotic* or *hypertonic* with respect to the surrounding water. Marine fish are *hypoosmotic* or *hypotonic* with respect to their medium.

Water enters the body of a freshwater fish via osmosis, always threatening to dilute the animal's body fluids and render physiological damage. Freshwater fish therefore never drink and excrete copious amounts of dilute urine to counteract the effects of osmosis. On the other hand, marine fish drink continuously and excrete only small amounts of concentrated urine in order to counteract dehydration. Constant ingestion of seawater makes marine fish more susceptible to water pollutants, a fact with important implications for the maintenance of marine aquariums. Low levels of toxic waste can have far-reaching consequences for the health of captive fish.

Marine invertebrates are *osmoconformers*, meaning that their body fluids are similar in composition to seawater. They freely exchange both water and solutes from their surroundings. Toxic substances are easily absorbed, resulting in extraordinary sensitivity. Levels of copper, for example, that are easily tolerated

by marine fish are rapidly fatal to many invertebrate groups.

Coral reefs do not occur randomly in the sea. The physical conditions in the ocean surrounding the reefs, not the chemistry of the water, determines coral distribution. These conditions limit the available habitat for the growth of the coral animals. Water clarity and exposure to sunlight, for example, are vitally important because of the essential symbiotic relationship that exists between corals and dinoflagellates. As photosynthetic one-celled organisms, dinoflagellates produce a portion of the coral's food and absorb some of its waste products. The temperature range under which this relationship is maintained is the primary reason coral reefs are restricted to the tropics, straying occasionally into the subtropics, as in the Florida Keys, but never quite achieving the splendor attained in warmer waters. Coral reproduction may be timed to the rhythms of tropical moonlight and tides. The distribution of species within the vicinity of a reef, and of the individual coral species that comprise the reef, may be determined by depth, currents, and wave action. Everywhere, dissolved oxygen at or near saturation supplies the needs of each living creature. For the aquarist, knowledge of physical parameters important to the reef's denizens becomes crucial to the design of the aquarium's life support systems: lighting, filtration, temperature control, and water movement. The chemistry of the seawater, however, remains unaltered among the various possible microhabitats that can be created by manipulating these design features.

In the last decade, aquarists have experimented with an array of methods for establishing and maintaining coral reef aquariums.

Technology and techniques associated with the reef-keeping hobby fit within several basic approaches or schools. Each of these has its own adherents and gurus. All would agree, however, that maintaining appropriate water conditions is of primary importance. Chemical analysis of the water remains the best method for determining if an aquarium system functions according to its design specifications.

Suggestions for lighting, filtration, temperature control, or other aspects of aquarium design should be sought from the books included in the reference list. Novices should be aware that no aquarium system can truthfully lay claim to being the best design. The systems described in any modern book on the subject of marine aquarium keeping can all be expected to perform as described by the authors. The amount of time you devote to aquarium maintenance and the care with which you choose the inhabitants of the tank (whether easily maintained or more demanding) will largely determine how easily you achieve success. Having a specific brand of filter or protein skimmer will, in the long run, make little difference. However, knowing if the system's performance is up to par is crucial. One way to determine this is to observe the behavior of the animals living in the aquarium. If they appear normal, are healthy, and (in the case of some invertebrates and seaweeds) are reproducing, you can assume that the filtration system is doing its job. A more precise way of evaluating the system is to carry out water tests on a regular basis and to compare the results with the target ranges recommended in Table 1 on page 81.

This book is a crash course in water chemistry. Subsequent chapters will be concerned only with the preparation of synthetic seawater and the measurement and control of certain of its chemical constituents. These include nitrogen, carbon dioxide, calcium, phosphorus, hydrogen ions, oxygen, and an assortment of substances lumped into the catchall categories of trace ions and dissolved organic carbon (DOC).

The author has endeavored to write with the needs of the average hobbyist in mind. Where technical terms are used, each is defined when it is first introduced and is included in the Glossary at the end of the book. The book is intended to serve both as a comprehensive guide to basic marine chemistry and as a quick reference when carrying out water tests as part of routine aquarium maintenance. Some of the information on the science of chemistry, especially that presented in the remainder of this chapter, may at first seem daunting to readers with little scientific background. Understanding this material is not essential to the successful maintenance of a marine aquarium. However, the better you can grasp these basic chemical concepts, the better you will be at evaluating the results of chemical tests you carry out and the more accurately you will assess problems and the means for solving them.

Basic Water Management

Stability, not only of the composition of seawater but also of the physical environment, characterizes coral reef habitats. The fish and invertebrates sold for marine aquariums are, for the most part, wild creatures that are collected from natural reefs. Because the reef environment undergoes such a narrow range of fluctuations with respect to biologically important parameters, reef dwellers have not evolved

Diadem Dottyback, **Pseudochromis diadema.**

Orange Spotted Filefish, **Oxymonacanthus longirostris.**

A snapping shrimp, **Alpheus bisincisu.**

the ability to cope with abrupt changes in their surroundings.

Freshwater fish, by contrast, may carry adaptation to environmental change to extremes. For example, one species of lungfish can lie dormant within a hardened ball of mud, feeding off body fat while in a state of suspended animation, to wait out the dry season. Freshwater fish have been maintained in captivity since ancient times, and the modern aquarium hobby is several decades old. Freshwater fish are regularly spawned in captivity. Numerous breeds exist, some of which are so greatly altered from their wild ancestors that they can survive only in the confines of an aquarium.

On the other hand, only about 10 percent of the organisms in the saltwater aquarium trade are produced through mariculture. The industry supplying cultured marine aquarium specimens remains in a fledgling condition compared with

Lionfish, **Pterois radiata.**

Angelfishes, including the Emperor Angel, **Pomacanthus imperator,** *require excellent water quality.*

Epibulus instidiator.

Foxface, **Lo vulpinus.**

hatchery production of freshwater species. Little opportunity has existed, therefore, for selective breeding of marine varieties that easily adapt to home aquariums.

Since marine fish lack the flexibility to adapt to fluctuating water conditions, the aquarist must ensure that the tank remains chemically stable. Herein lies the essence of marine water quality management. Identify changing trends, and take steps to restore normal conditions before fluctuations become too large. This is the test and tweak approach. Test the water periodically. If anything is found to be straying from the target value, tweak it back into line.

Adjusting some physical parameters is relatively simple. Small increases or decreases in temperature can be achieved by altering the thermostat setting, for example, or a timer can be reset to change the day length. Correcting a problem with water chemistry may be more complicated. More than one factor may be the cause or several factors may be working in concert. This makes diagnosing a problem tricky, especially for an inexperienced aquarist. Without routine testing, trouble often goes unnoticed until the deviation from operating norms becomes extreme. By then, significant damage will have been done to the biological processes that must remain in balance in order to maintain proper water conditions. Effecting change, altering the pH for example, should always be done gradually, even though one may be correcting a problem. Beginners should remember that problems with water chemistry usually happen quickly, while improvements usually take place slowly. Routine monitoring offers the best insurance against water quality deviations that might affect fish health.

Marine fish react to improper water conditions with varying levels of stress. If a fish remains unrelieved of its stress for too long a period, it often succumbs to an infestation of parasites. That, in turn, may lead to other pathological conditions, causing the fish's health to decline rapidly. Without appropriate care, and sometimes in spite of it, the fish may die. Such a cascade of events may also include the spread of disease to other fish in the tank. The least-hardy individuals become vectors for parasites or bacteria. Even the most durable specimens may be unable to resist the combined onslaught of deteriorating water conditions and active disease in their tank mates. Ultimately, the tank crashes, and most or all of its inhabitants die. Such a scenario often results from neglecting water tests or from inadequately interpreting test results. Rarely, trouble can come from misadjusted test equipment or contaminated chemicals. However,

negligent or uninformed aquarists more frequently find themselves at fault.

When the aquarist adopts a routine testing program and keeps a record of the observations made, changes in the condition of the water over time will soon become evident. Early in the life of the aquarium, within the first few months after it has been set up, various biological processes have not yet reached equilibrium. Test readings will indicate seemingly random changes in water parameters over the course of days or weeks. In an older, established aquarium, changes typically occur more gradually and with a greater degree of predictability. Some degree of deviation from ideal conditions will, however, inevitably occur. For example, as stony corals remove calcium from the water, replenishment becomes necessary. This can be accomplished either through the natural dissolution of calcium-containing rock or sand used to construct the artificial reef or by the addition of calcium ions from an external source. The aquarist should prepare to manage such anticipated changes and must also be prepared to cope with unanticipated changes. These include a sudden, sharp increase in dissolved organic carbon resulting from an unnoticed animal death or a decrease in dissolved oxygen resulting from equipment failure.

What Is Water Quality?

In aquarium books and magazines, one frequently encounters the term *water quality*. Authors exhort beginners to "maintain good water quality" or warn that a particular species "demands perfect water quality." Water quality may be defined in terms of the biological needs of the inhabitants of the aquarium.

If the physical and chemical parameters of the aquarium water are within acceptable ranges (as observed in natural habitats) for the community of species housed in the tank, then the aquarium may be said to have appropriate water quality. Target parameters are therefore chosen based upon knowledge of the inhabitants and their needs. For a coral reef aquarium, these parameters can be precisely defined.

Determining the *carrying capacity* of a given aquarium, that is, estimating how many fish may be successfully maintained, has always presented a thorny problem. Many variables, ranging from the design of the filtration system to the metabolism of individual species of fish, interact, making calculations complicated. Most dealers and hatchery managers and also many seasoned hobbyists judge the carrying capacity of a system by making educated guesses based on prior experience. Such difficulty in predicting carrying capacity leads most authors to advise beginners to start with a larger system and add animals gradually.

On the other hand, determining if a particular system is operating below its carrying capacity at a given moment is a simple matter of carrying out tests for nitrogen compounds, pH, dissolved oxygen, and carbon dioxide. Comparing these values to recommended norms provides an obvious answer. If wastes are not accumulating and sufficient oxygen is available, the system is operating at or below its maximum capacity. Acceptable ranges for marine water chemistry are presented in Table 1 on page 81.

Three major biological activities operate within an established aquarium. These are *nitrogen cycling*, *photosynthesis*, and *food web* formation. Nitrogen cycling is the exchange of compounds containing nitrogen among organisms in a food web. It is facilitated by bacterial activity as well as the metabolism of higher organisms. Photosynthesis is the process by which plants and algae use sunlight to form food molecules from carbon dioxide and water. A food web is a community of photosynthetic and nonphotosynthetic organisms that are ecologically related to each other as prey and predators.

These three dynamic processes constantly alter water chemistry in the closed system of an aquarium. When an equilibrium among these processes is achieved, fluctuations in water chemistry remain relatively minor over the course of time. Regular maintenance suffices to maintain the stability of the system. When the aquarium is operating below its carrying capacity, chemical tests performed at regular intervals would show, for example, that the amount of nitrate remains relatively constant. If the carrying capacity of a system is exceeded, toxic waste produced by the inhabitants, such as carbon dioxide and ammonia, typically accumulate rapidly.

Basic Chemistry

Marine aquarium keeping demands a broader understanding of water chemistry than does freshwater aquarium keeping. Freshwater is not, of course, pure water. It varies in the amount of dissolved minerals and organic matter it contains and in the resulting acidity or alkalinity of the solution that forms. This book will therefore launch its study of marine water chemistry by first examining the chemistry of freshwater habitats. Before beginning, however, important terms and concepts must be defined

Even under ideal conditions, some species can be challenging (sea slug, **Chromodoris bullocki***).*

because they will be repeatedly referred to throughout the remainder of the book.

Everything that exists is either matter or energy. Albert Einstein proved that these two concepts are really just different aspects of the same thing. Chemistry concerns itself with the interactions between different types of matter, involving, on a cosmic scale, relatively small amounts of energy. Various branches of

Hardy species like the Blacktail Humbug, **Dascyllus melanurus,** *tolerate average water quality.*

Oceanic Seahorse, **Hippocampus kuda.**

chemistry devote attention to specific types of interactions or specific types of matter. Physical chemistry, for example, is concerned with the transfers of energy that occur during chemical reactions. Organic chemistry, in contrast, focuses upon the single element, carbon, and the multitude of compounds in which it occurs. For the purposes of this manual, aquarium chemistry is the investigation by means of

Flame Hawkfish, **Neocirrhites armatus.**

the oxygen and nitrogen in the atmosphere and metals such as iron, gold, and copper.) Atoms, as most people know, are themselves comprised of smaller particles. Electrons, bearing a negative electrical charge, orbit around the atomic nucleus, which is comprised of positively charged protons and uncharged neutrons. The combined mass of the protons and neutrons results in a characteristic *atomic weight* for each type of nucleus. Because the number of protons and electrons in a given type of atom is always equal, atoms have no net electrical charge. Protons and neutrons are

Indian Ocean Cleaner Shrimp, **Stenopus zanzibaricus.**

chemical analysis of the changes that occur in water over the life of an aquarium system.

   Particles and bonds: All matter is composed of *atoms.* An individual atom is the smallest unit of a *chemical element* that retains the properties of that element. (A chemical element is a substance that cannot be reduced to simpler substances through ordinary chemical processes. Familiar examples of elements are

Clown Triggerfish, **Balistoides conspicillum.**

held within the atomic nucleus by forces that are the subject matter of nuclear physics and are not in the scope of this discussion. The behavior of the electron, however, figures in all of the chemistry that will be presented.

The electrons arranged around the nucleus of the atom can exist in various energy states, some of which are inherently more stable than others. The outermost electrons typically exist in a state of energy instability. However, interactions between atoms can alter that state, achieving a more stable one. One can think of this as if the electrons were trying to achieve a stable orbit around the nucleus. This tendency of electrons is the basis for the coming together of chemical elements to form *compounds*, substances formed from two or more elements that exhibit properties different from those of the elements individually. In a *chemical reaction*, two or more atoms combine to form a molecule. A molecule of oxygen (O_2), for example, is comprised of two oxygen atoms bound together. A molecule of water (H_2O), similarly, results when two atoms of hydrogen combine with one atom of oxygen. The construction of a simple molecule can usually be deduced by examining its formula, as the foregoing examples demonstrate. The force holding the atoms together in a molecule is known as a *chemical bond*.

In some chemical reactions, electrons are actually transferred from one atom to the other. In the process, the atom that gives up an electron, having lost one of its negative charges, retains a positive charge. The other atom acquires a negative charge because it now has one more electron than it does protons. Such charged atoms, in which the number of electrons differs from the number of

protons in the nucleus, are called *ions*. Table salt, sodium chloride (NaCl), is a common example of a molecule formed when a positive and a negative ion come together. The negative charge on the chloride ion is attracted to the positive charge on the sodium ion, holding the two together.

The behavior of ions: The chemistry of seawater primarily involves the study of ions. When compounds such as sodium chloride are dissolved in water, the individual molecules separate into their constituent ions. Each molecule produces an equivalent number of positively and negatively charged ions, and the resulting solution is electrically neutral. Although several ions are major constituents of seawater, certain ions of great interest to aquarists typically occur in low amounts.

Acids and bases: A particularly important ion is the one produced when hydrogen gives up its single electron to become a free proton, H^+. Pure water, H_2O, dissociates into a hydrogen ion, H^+, and a hydroxyl ion, OH^-. Chemical compounds that are hydrogen ion donors are known as *acids*, while compounds that are hydrogen ion acceptors (or hydroxyl ion donors) are *bases*. Bases typically form solutions that are alkaline. The degree of acidity or alkalinity of a solution such as seawater has important implications for the biological activities of organisms inhabiting it. This condition of a solution is known as its *pH*. It influences so many important processes in the aquarium that considerable attention will be later devoted to its measurement and interpretation (see page 51).

Redox reactions: Some elements readily give up their electrons in chemical reactions. Others are more reluctant to do so and are, in fact, more likely to be acceptors of electrons. (Note that this dicussion concerns the exchange of electrons, not protons.) Oxygen, for example, is so greedy with its electrons that it never donates one under normal circumstances but rather acquires two, becoming the oxide ion, O^{2-}. The tendency to acquire electrons is known as *electronegativity*. Chlorine is another element that is strongly electronegative, forming the Cl^- ion, chloride. Chlorine and oxygen, as the recipients of the electrons, are called *oxidizing agents*. When they acquire an electron from a donor atom, they are said to be *reduced*. (You can remember this if you think of negatively charged electrons reducing the charge on the atom from 0 to –1.) The donor atom, on the other hand, loses electrons and is said to become *oxidized* in the process. The donor of the electrons is known as the *reducing agent*.

Exchanges of electrons do not always shift the charge on the ion from positive to negative or vice versa but rather may simply alter the magnitude of the charge. Nitrogen and oxygen can team up, for example, to form either the nitrate ion, NO_3^-, or the nitrite ion, NO_2^-. Note that although an atom of oxygen is lost in the change, the charge on the ion remains –1. This is an example of a *redox reaction*. The complete reaction is written thus:

$$NO_2^- + O_2 + 2H^+ \rightarrow NO_3^- + H_2O$$
(equation 1)

In this equation, nitrite is oxidized to nitrate, and oxygen is reduced to water. The oxidizing agent is oxygen, and the reducing agent is nitrite.

The tendency for redox reactions to occur in a solution is the *redox potential* of that solution and is expressed in millivolts (mV). The redox potential of a complex solution such as seawater is the sum of all the individual redox potentials present. In the aquarium hobbyist literature, redox potential is often referred to as *oxidation-reduction potential*, abbreviated ORP. This parameter can be measured electronically. Systems in which oxidizing reactions are favored have a more positive redox potential, while those in which such reactions are less likely to occur have a more negative redox potential. Some aquarists place great faith in measurements of ORP, regarding it as the best way to define the overall health of the system with a single number. A redox potential of around 400 mV is generally recommended. However, one should be aware that this value fluctuates with pH. Seldom is the pH reported alongside the ORP value in the aquarium hobbyist literature, although anecdotal evidence suggests that reported readings are obtained in aquariums at a pH of 8.2 to 8.3, the value for natural seawater.

Redox potential can be increased by adding oxidizing agents such as hydrogen peroxide or, more commonly, ozone. Ozone, O_3, can be produced by means of a device known as an ozone generator. Coupling an ozone generator to a voltmeter permits automated control of the redox potential. One should be aware, however, of the pitfalls associated with ORP measurement. The primary difficulties involve calibration and the need for frequent replacement of the sensing probe. Standardized solutions that can be used for calibration of an ORP probe must be made up immediately before use, because they degrade rapidly upon storage. Many such standards, furthermore, are comprised of toxic or otherwise dangerous chemicals that may pose a hazard if used by inexperienced persons in a

Blackcap Basslet, **Gramma melacara.**

Reef Lobster, **Enoplometopus debelius.**

Anemone Crab, **Neopterolisthes maculatus.**

Combtooth Blenny, **Astrosalaria fuscus.**

household setting. Adding to the problem is the fact that ORP probes have a relatively short life span. It is possibly shortened further by the necessity for frequent cleaning to remove growths of algal and bacterial films that inevitably coat the probes while immersed in the aquarium.

Not all aquarists agree regarding the necessity for monitoring and/or controlling the redox potential of a marine aquarium. Certainly, aquariums can be maintained successfully without resorting to the use of this technology.

Pajama Cardinalfish, Sphaeramia nematopterus.

Blue-Legged Hermit Crab, Clibanarius tricolor.

Pristigenys niphonia.

Yellownose Goby, Stenogobiops xanthorhinia.

True, many hobbyists consider redox potential a reflection of the overall health of the system, but such judgments are subjective. Too often, the author has seen hobbyists proud to report an ORP of 425 mV when the appearance of the organisms in the tank clearly indicated something was seriously wrong. Like any other parameter, ORP cannot be relied upon as the sole indicator of the state of the aquarium environment. ORP readings must always be evaluated in the context of other measurements, such as pH, dissolved oxygen, and the concentration of

Bangaii Cardinalfish, Pterapogon kauderni.

nitrogenous wastes. Since the latter may be measured more simply and at less expense, the value of ORP measurements for the novice marine hobbyist remains doubtful.

Covalent bonding: Two atoms may share a pair of electrons between them, thus producing a *covalent bond*. Covalent bonding is responsible for producing *complex ions* comprised of several atoms, such as nitrate, NO_3^-. Complex ions react with other ions as a unit. Complex ions important in marine aquarium management include nitrate, nitrite, ammonium, carbonate, bicarbonate, phosphate, and the host of largely unidentified organic ones that collectively make up dissolved organic carbon (DOC).

The properties of water: The bonds between the hydrogen atoms of a water molecule and their oxygen atom are covalent, but the shared electrons tend to spend more of their time near the oxygen nucleus than they do near the hydrogen nuclei. This results in the molecule's having a positive end and a negative end. Therefore, the water molecule is *polar*. The polarity of the water molecule, and the unique properties thus conferred upon water, have far-reaching implications for all of biology.

The negative pole of a water molecule is attracted to the positive pole of its neighbor. This weak interaction, known as a *hydrogen bond,* is responsible for the fact that ice is less dense than liquid water. Were that not the case, the oceans would all be solid ice by now. Polar ice would constantly sink to the bottom, and the upwelling surface water would be subsequently frozen. Water expands when it freezes, because hydrogen bonds cause ice crystals to form a latticework structure with more space between the molecules than in liquid water. Ice therefore floats and can move

around on the ocean's surface to be melted again by the Sun's heat. Consequently, life can exist on the planet.

Because hydrogen bonds tend to hold water molecules in place with respect to each other, a lot of energy is needed to get them moving, that is, to raise the temperature. The movement of molecules as heat is measured in units known as *calories* (cal). *Specific heat* is the amount of energy in calories required to raise the temperature of 1 gram (1 g) of a substance by 1 degree Celsius (1°C). Water has the highest specific heat (1.00 cal/g/°C) of any commonplace substance. Conversely, the calorie is defined as the amount of heat needed to raise the temperature of 1 gram of water by 1 degree Celsius. (Note: 1 dietary Calorie is equivalent to 1,000 heat calories.) Owing to water's high specific heat, aquatic habitats are thermally stable, protecting the life they harbor from the wide temperature swings to which terrestrial species must adapt. This is one reason for the relatively narrow range of temperatures within which coral reefs develop.

Water's polarity has another effect. When chemical compounds are placed into water, the negative ends of the water molecules tug at the positively charged ions in the compound. The positive ends of the water molecules behave similarly toward the compound's negatively charged ions. This helps to pull the compound apart, making water a nearly universal solvent. The universal solvent effect transforms almost pure water that falls to the ground as rain into the solution that eventually melds with the sea at the mouth of a great river. Eons of rainfall and river flow have stripped the surface rocks of the planet of soluble compounds, filling the ocean basins with seawater. The composition and behavior of this unique substance, and the preparation of an artificial substitute for aquarium use, are the focus of the next chapter.

Measurements and units: Much of the language of chemistry has to do with units of measurement. The internationally accepted standard for chemical measurements calls for the use of the metric system, and only occasionally will this book refer to gallons or pounds in discussing aquarium chemistry. Instead, grams (g), kilograms (kg; 1,000 g = 1 kg), liters (L), milliliters (mL; 1,000 mL = 1 L) and degrees Celsius (°C) will be used when carrying out aquarium water analysis. The concentrations of various ions will usually be expressed as the mass of the ion in milligrams per liter of water (mg/L). Aquarium references may also refer to parts per million (ppm), which is a slightly different number than milligrams per liter but close enough for practical purposes. The concentration of a substance in parts per million in seawater equals the concentration in milligrams per liter, multiplied by the specific gravity. The *specific gravity* of a solution is the ratio of the weight of a given volume of that solution to the weight of an equal volume of pure water. For a sample of aquarium water, the specific gravity might be 1.0260. Therefore, if the concentration of calcium in this sample is 300 mg/L, its concentration in parts per million is $300 \times 1.0260 =$ 307.8 ppm. This is a trivial difference for aquarium maintenance purposes. Aquarium books typically use the terms ppm and mg/L interchangeably. The measurement of specific gravity is useful to the aquarist in estimating the salinity of the water, as will be explained in detail in a subsequent chapter.

How water tests actually work: If two compounds are introduced into the same solution, their component ions will interact in a predictable way. This fact is the basis for all the types of chemical water tests that one might wish to perform on aquarium water. To illustrate how this works, consider the simple example of adding carbon dioxide, CO_2, to a solution of limewater, $Ca(OH)_2$. The ions recombine to form calcium carbonate, $CaCO_3$, and water, H_2O, according to the following reaction:

$$CO_2 + Ca(OH)_2 \rightarrow CaCO_3 + H_2O$$
(equation 2)

Note that the total number of atoms on the left side of the equation equals the total number of atoms on the right side. There are four atoms of oxygen, two of hydrogen, and one each of carbon and calcium. Only the arrangement of the atoms is different after the reaction has taken place. All chemical reactions behave this way. The atoms recombine in specific ratios, with none left over after the reaction is complete. This principle was discovered by the Italian ecclesiastical lawyer and natural philosopher Amedeo Avogadro (1776–1856).

Avogadro found that the atomic weights of the individual atoms in a molecule can be added and then an equal number of grams of that compound can be weighed out to arrive at a gram-molecular weight, or *mole*, of the compound. The atomic weight of oxygen is 16, of hydrogen is 1, and of calcium is 40. Two oxygen atoms and two hydrogen atoms plus one calcium atom equal 74 for the molecular weight of calcium hydroxide. Therefore, 1 mole of calcium hydroxide weighs 74 grams.

Why is this information of use in chemical analysis? Because 1 mole of any substance contains the same number of elementary

Left: Potter's Angelfish,
Centropyge potteri.

Below left: A sea slug,
Hypselodoris edenticulata.

Below top: Leaf Fish,
Taenianotus triacanthus.

*Below bottom: Whitehurst's
Jawfish,* **Opisthognathus
whitehurstii.**

Green Chromis, Chromis viridis.

Mertens' Butterflyfish, Chaetodon mertensii.

Liomena.

Red Sea Angelfish, Pomacanthus maculosus.

Sharpnosed Goby, Gobiosoma evelynae.

particles. This amount is known as Avogadro's number. This is a huge number, 602,000,000,000,000,000,000,000 particles, or in scientific notation, 6.02×10^{23}. The number itself is not important. What is significant is that 1 mole of anything contains this same number of particles. In a chemical reaction, each particle of one reactant combines with a particle of the other reactant. Therefore, if it is known how many particles are present of substance A, how many particles of substance B will be used in a reaction between A and B can be predicted. Since 1 mole of any substance contains the same number of particles, chemical reactions occur in molar ratios. Knowing how many moles is in a particular solution of calcium hydroxide, for example, enables the aquarist to predict how many moles of carbon dioxide can be expected to react with it completely.

Adding 1 mole of any compound to 1 liter of water produces a 1 molar (1 M) solution of that compound. When two solutions are combined, every unit (such as a drop, milliliter, or teaspoonful) of a 1 molar solution of one reagent will react completely with a precisely equal unit of a 1 molar solution of the other reagent. Furthermore, since moles can be easily converted to grams or milligrams, the actual concentration of an unknown substance in a sample can be extrapolated by observing how much reactant of a known molar concentration is needed to react completely with the sample. Many of the chemical tests performed on aquarium water take advantage of this convenient mathematical relationship.

For instance, what if the aquarist would like to know how much carbon dioxide might be present in a seawater sample? By inspecting equation 2 one can see that 1 molecule of carbon dioxide combines with exactly 1 molecule of calcium hydroxide. Carbon's atomic weight is 12, and, as noted, the atomic weight of oxygen is 16. Thus, the molecular weight of carbon dioxide is 12 plus 32 ($12 + 2 \cdot 16$), or 44. Therefore, 44 grams of carbon dioxide will combine completely with 74 grams ($40 + 2 [16 + 1]$) of calcium hydroxide. Assume that the test reagent contains 7.4 grams of calcium hydroxide per liter, that is, it is a 0.1 molar solution. Suppose the aquarist removes 1 liter of water from the aquarium and finds that it must be combined with an equal amount of the test reagent to achieve a complete reaction. (Do not worry about how the aquarist determines that the reaction is complete. The important point is the combining ratio.) Since the combining ratio is 0.1 mole (7.4 grams) of calcium hydroxide to 0.1 mole (4.4 grams) of carbon dioxide, the 1 liter of sample must contain 4.4 grams of carbon dioxide. The concentration of carbon dioxide is therefore calculated to be 4.4 grams per liter (gm/L).

Neutralization reactions and acid–base testing: Reactions among acids and bases are of particular interest to chemists and aquarists alike. Recall from the earlier discussion that acids are hydrogen ion donors. Strong acids, such as the hydrochloric acid (HCl) used in swimming pool maintenance, dissociate into their component ions almost completely when they are dissolved in water.

$$HCl \rightleftharpoons H^+ + Cl^-$$

(equation 3)

By inspecting this equation, 1 mole of hydrochloric acid will obviously produce 1 mole each of hydrogen ions, H^+, and chloride ions, Cl^-. Using 1 mole of another strong acid, sulfuric acid (H_2SO_4), produces 2 moles of hydrogen ions and 1 mole of sulfate ions, SO_4^{2-}, when it dissociates:

$$H_2SO_4 \rightleftharpoons 2H^+ + SO_4^{2-}$$

(equation 4)

Acids and bases combine with, or neutralize, each other in exact molar ratios. Therefore, 1 mole of acid neutralizes 1 mole of base. (The term *neutralization* is used to describe these reactions because the resulting products, water and another compound, produce a solution neither acidic nor basic. The resulting solution is chemically neutral.) Therefore, when combined with a base, 1 mole of hydrochloric acid will combine with 1 mole of base, and 1 mole of sulfuric acid will combine with 2 moles of base. Chemists say that 1 *equivalent* of base is neutralized by a 1 molar solution of HCl, while a similar solution of H_2SO_4 will neutralize 2 equivalents of base.

$$HCl + NaOH \rightleftharpoons H^+ + Cl^- + Na^+ + OH^- \rightleftharpoons NaCl + H_2O$$

(equation 5)

$$H_2SO_4 + 2NaOH \rightleftharpoons 2H^+ + SO_4^{2-} + 2Na^+ + 2OH^- \rightleftharpoons Na_2SO_4 + 2H_2O$$

(equation 6)

A solution containing 1 equivalent of acid or base is called a *normal* solution. A given amount of a 1 normal (1 N) solution of any acid will neutralize exactly the same amount of a 1 normal solution of a base. This principle can be used to measure the amount of base in an unknown sample by determining how much acid of known normality is needed to neutralize the sample completely. This is known as a *titration* and is the basis for several types of aquarium water tests. Later, the specifics regarding titration of seawater will be discussed to understand acid-base balance in the aquarium better. The important point in this discussion is that neutralization reactions occur in predictable ratios. Knowing the normality of one reactant and the amount needed to combine with an unknown in a water sample allows the aquarist to calculate the concentration of the unknown.

Equilibrium reactions: Note that the arrows in equations 3, 4, 5, and 6 point in both directions. This is one way of indicating an *equilibrium* reaction, one that can proceed in either direction. Not only can molecules dissociate into individual ions when dissolved in water, a proportion of those ions can recombine to form the original molecule. In the case of HCl, dissociation is strongly favored over recombination, and the reaction is shifted nearly 100 percent to the right. HCl is therefore a *strong* acid. In the case of *weak* acids, by contrast, the tendency to dissociate is less. The reaction may be shifted to the right by considerably less than 100 percent. This is the case with carbonic acid, H_2CO_3, formed when carbon dioxide is dissolved in water. Equilibrium reactions, especially those involving carbon dioxide and carbonic acid, are important factors in determining the acid-base balance of seawater. A more complete explanation of these reactions appears in the discussion of aquarium pH measurement and control, beginning on page 51.

SEAWATER

The oceans cover four-fifths of Earth's surface. Despite the vastness of this realm, seawater is remarkably similar in chemical composition throughout the world. Understanding the special properties of seawater is an important skill for marine aquarium keeping.

Water Hardness and Alkalinity

Aquarists who have some experience with freshwater tanks know that only two chemical parameters are typically cited to define specific freshwater habitats, pH and *hardness* (the amount of dissolved substances the water contains). They may also know that the concentrations of nitrogenous compounds—nitrate, nitrite, and ammonium—must be controlled because these are toxic to the fish. If living plants are growing in the aquarium, carbon dioxide and possibly phosphate may be of interest. Providing a sufficient concentration of oxygen for the fish's respiration may also be a concern, although this is rarely a problem in a reasonably stocked tank. Marine aquarium hobbyists need to concern themselves with all of these parameters and a few more as well.

An aquarium for coral reef species requires a supply of seawater, either natural or synthetic. (Anemone crab and anemone.)

The pH of freshwater habitats can range from acidic (pH 5.0) in a rain forest stream to strongly alkaline (pH 8.2) in an African rift lake. Seawater, however, always remains at a pH of about 8.2. Unless steps are taken to counterbalance fluctuations, the pH of an aquarium may vary somewhat over the course of a day without apparent harm to the tank's inhabitants.

Maintaining a stable pH is accomplished because of seawater's *buffer system*, or its ability to resist a change in pH when acid is added. This capacity is known as *alkalinity*. Measuring alkalinity is of greater importance to marine aquarists than to their freshwater counterparts. Other names for this parameter are *carbonate hardness* and its German equivalent, KH. In freshwater systems, the term *hardness* is simply used.

Freshwater acquires hardness when minerals, especially calcium and magnesium carbonates, are dissolved by water as it flows over Earth's surface. These compounds are only slightly soluble in water. A freshwater aquarium's total hardness values rarely exceeds 400 mg/L. Seawater, on the other hand, contains most of the soluble salts once present in Earth's crust and is 100 times harder than freshwater, at 35,000 mg/L. Seawater chemists, therefore, use *salinity*, expressed in grams per liter (gm/L) or parts per thousand (ppt, symbolized as ‰) to describe the total weight of dissolved ions in seawater. Full-strength seawater therefore has a salinity of 35 ppt. The measurement of salinity will be given special consideration at the end of this chapter.

Much of the hardness measured in freshwater is due to dissolved calcium compounds. The calcium concentration of seawater, because of its importance to the development of stony coral skeletons, is another parameter that should be measured on a routine basis if corals are being maintained. Alkalinity, hardness, and calcium concentration are all determined by the same method, titration, in both freshwater and seawater aquariums.

Other dissolved compounds: Nitrogenous compounds concern marine aquarists as well as their freshwater counterparts, although the levels of toxicity are different. Phosphate, carbon dioxide, and dissolved oxygen all play the same roles, albeit at different concentrations, in both freshwater and marine habitats. Seawater chemistry, therefore, should not be completely unfamiliar to readers who have experience only with freshwater systems. The much greater amounts of dissolved substances in seawater do, however, present special problems for aquarium maintenance operations. These problems involve not only chemical analysis but also seawater storage and the formulation of synthetic seawater.

Natural Seawater

Rainfall picks up minerals from Earth's crust and eventually merges with the ocean, adding its components to the complex mixture called seawater. Slow-moving biogeochemical cycles counterbalance both the inputs of minerals and the evaporation of water, maintaining the composition of seawater with remarkable constancy. Although all chemical elements and a host of other substances, such as pollutants, can be identified in seawater, a large proportion of the mineral content is sodium chloride, common table salt. Away from the diluting influence of an incoming river, the concentration of dissolved minerals in seawater is 3.5 percent.

Seawater is composed of 11 major ions. These constituents make up 99.9 percent of the dissolved substances in seawater. They are chloride, sulfate, bicarbonate, bromide, borate, fluoride, sodium, magnesium, calcium, potassium, and strontium. These major ions are said to be *conserved*. This means that although the total amount of these ions can vary locally, that is, salinity may be different in different locations, the relative proportions of these ions remain constant. This fact is very important. Marine organisms such as corals are adapted to a specific set of chemical conditions in the seawater that surrounds them. This explains why maintaining any marine aquarium involves testing and adjusting the chemical conditions of the water in the tank. Besides the major ions, all of which are present at a concentration of 1 part per million (ppm) or greater, two other groups of ions are found in seawater. Minor ions are those that are present at a level greater than 1 part per billion (ppb) but less than 1 ppm. Trace ions are those present in concentrations less than 1 ppb.

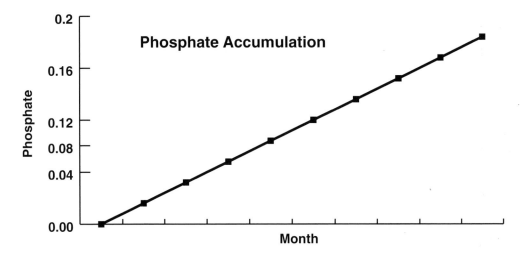

This graph shows the result of using tap water containing 0.10 ppm of phosphate to perform monthly, 20 percent water changes in a hypothetical marine aquarium. Assuming there are no additional inputs of phosphate (an unlikely scenario in a real aquarium system) the tank will accumulate more phosphate than is present in the tap water within seven months. An algae bloom would probably occur when phosphate rises above 0.05 ppm, at around four months.

Minor variations upon the principle of conserved elements occur in the unusually biologically active vicinity of a coral reef. Calcium, for example, may be present in a slightly lower proportion there than in deeper water. Extraction of calcium by reef-building corals is responsible for this local discrepancy. The calcium eventually returns to the sea, however, when the coral dies and the skeleton erodes, dissolving once again into the water. The calcium concentration of aquarium water should be routinely monitored when maintaining stony corals or any of the many other organisms that absorb this element.

The major ions of seawater account for most of the salinity but are of little interest to the aquarist individually. Only by virtue of the interactions among them is their importance realized. For example, bicarbonate, (HCO_3^-), a participant in the biological formation of skeletal calcium carbonate, is one form of carbon present in seawater. The total concentration of bicarbonate is about 28 ppm. The other forms of carbon in seawater are carbonate and carbon dioxide. These substances are largely responsible for the buffering effect in seawater. This buffering effect causes the pH to remain relatively constant despite dynamic biological processes that either remove ions from solution (calcification by corals) or add more (respiration by all living organisms). The precise concentration of bicarbonate is not, by itself, a concern in maintaining the aquarium. Keeping the buffering system intact, however, is of considerable importance.

Cuttlefish, Sepia officinalis, *are adapted to chemical conditions found in the Mediterranean Sea.*

The Swissguard Basslet, Liopropoma rubre, *lives in deep waters of the Caribbean.*

A sea slug, Chromodoris innulata.

Sea Apple, Pseudochlorochirus violaceus.

Squareblock Anthias, Pseudanthias pleurotaenia.

Yellow Sea Cucumber, Cucumaria miniata.

A dragonet, Synchiropus *sp.*

Clark's Anemonefish, Amphiprion clarkii.

Apart from calcium, none of the major constituents of seawater need be monitored in the course of routine aquarium maintenance. Of far greater interest are substances that are present in relatively tiny amounts, as will be elucidated in the chapter "Dissolved Components."

Synthetic Seawater

The following sections discuss the use of synthetic seawater in the marine aquarium. Several aspects, including the water used, salt mixes, preparation, and storage, will be discussed.

Water Itself

Suitable synthetic seawater for a variety of needs can be prepared using water straight from the tap in most locations. However, impurities, even in cities with high standards for drinking water, may cause problems when tap water is used for a display aquarium. One reason for this is the accumulation of small amounts of impurities over time. When maintaining a typical marine aquarium, one should change about 20 percent of the water monthly. If, for example, the tank holds 50 gallons (200 L), each month 10 gallons (40 L) should be siphoned out and discarded. Then, 10 gallons (40 L) of properly prepared seawater should be used for replenishment.

An aquarist should understand that any substance present in the replacement seawater will gradually accumulate in the aquarium, as can be seen when the concentration of that substance is plotted with respect to time on a graph. Phosphate, for example, is a frequent contaminant of municipal tap water. As it accumulates in the aquarium, it provides fertil-

izer for *blooms*, which are sudden, rampant growths of algae that interfere with the aesthetics of the display. Using demineralized water when preparing synthetic seawater avoids introducing impurities.

Obtaining a supply of demineralized water may be accomplished by several methods. Water may be purchased; distilled water is sold in bottles in any grocery store. However, this may prove to be too expensive and cumbersome a way to meet the requirements of a large marine system. For in-home production of useful quantities of demineralized water, one basically has three options—deionization (DI), reverse osmosis (RO), or a combination of the two. Each method has advantages and disadvantages.

The deionization process removes tap water contaminants by means of specially formulated chemical resins over which the water flows. *Ion-exchange resins* swap ions of sodium for the ions removed from the water supply. Depending upon the formulation, resins may absorb positively charged *cations* such as calcium or negatively charged *anions* such as phosphate. To effect removal of a broad spectrum of contaminants, therefore, two types of resins must be combined in the treatment vessel. Because the ion-holding capacity of each resin particle is eventually exhausted, the resins must be periodically recharged. This is a relatively simple procedure. However, it involves the use of hazardous chemicals and should not be carried out in a home environment by untrained persons. The aquarist therefore must either use a small, disposable DI unit or have a larger system installed and regularly serviced by a water treatment company. The supply of DI water is available on demand each time the tap is opened, a major advantage. High initial cost and the need for recharging the resins present the main disadvantages of DI systems.

Reverse osmosis is perhaps the most widely used method of home water purification. The units may often be found in do-it-yourself centers in areas with hard water. Also, mail-order suppliers of RO systems abound. Although RO water is not produced on demand, the low cost, ease of operation, and simple maintenance requirements are decided advantages. RO systems are easily scalable, with units available that produce from 15 gallons (60 L) to over 100 gallons (400 L) of demineralized water daily. Line pressure forces water through a semipermeable membrane, which in effect strains out contaminant molecules. Impurities are discharged in a stream of wastewater that may be used for laundry or irrigation purposes, while demineralized water is released into a reservoir. Most units come with prefilters for particulate matter and chlorine removal. Prefiltration increases the useful life of the membrane, which typically must be replaced after several thousand gallons (liters) of product water have been made. In areas where tap water is expensive or water conservation is a concern, a major disadvantage of RO is the production of wastewater, about 5 gallons (20 L) for every 1 gallon (4 L) of product water.

Reverse osmosis by itself can reduce contaminant levels from about 90 percent to over 98 percent, depending upon various factors such as the temperature and pressure of the water supply, initial contaminant levels, and so forth. If exceptionally high contaminant levels are present in the water supply, purification by RO alone may not reduce contaminant levels

below the threshold desired for aquarium use. In this case, the output of the RO unit can be passed through a DI unit for further purification. Since impurities have already been significantly reduced by the RO unit, the DI resins will need less frequent replacement than would have been the case if they had been used alone.

Obtaining an official analysis report for your tap water from your utility company is helpful. Depending upon where you live, it might even be free of charge. Check the numbers for substances that are factors in aquarium maintenance—phosphate, nitrate, and metals such as iron. Of course, none of these will be present in amounts considered harmful for drinking purposes. However, the tap water in the author's community, for example, contains over 2,000 times the phosphate content of the seawater

TIP

Evaporation

One source of change in the chemistry of marine aquarium water, evaporation, is frequently overlooked. Do not ignore it, however. Only water evaporates from the tank, leaving behind the dissolved salts. As a result, the salinity increases. Adding seawater as a replacement would only compound this effect. Adding demineralized water facilitates maintenance of the salinity, because no dissolved salts are added along with it. This is another reason aquarists should provide themselves with a constant supply of demineralized water.

Chambered Nautilus, **Nautilus pompilius.**

Yellowbar Angelfish, **Pomacanthus maculosus.**

Goldentail Moray, **Gymnothorax miliaris.**

surrounding a coral reef. After inspecting the report, decide if a 90 percent reduction will bring any problem compounds within acceptable range. If so, an RO unit will likely meet your water purification needs. If a significant amount of any contaminant remains after a 90 percent reduction, the RO water should then be treated with a DI unit.

Both RO and DI systems should have prefilters that remove particulate matter, such as mud and sand. The housing for this filter, supplied as a cylindrical cartridge, should be transparent to facilitate inspection. A mud and sand filter should be changed when it is visibly dirty. The time interval between changes will depend upon the condition of your tap water; once or twice annually is typical. The membrane of most RO units should be protected from the chlorine that is added to municipal water as a disinfectant. An activated carbon prefilter should be installed and periodically replaced according to the manufacturer's directions,

Starck's Damselfish, Chrysiptera starcki.

Blood Red Coral Shrimp, Lysmata debelius.

Pinnatus Batfish, Platax pinnatus.

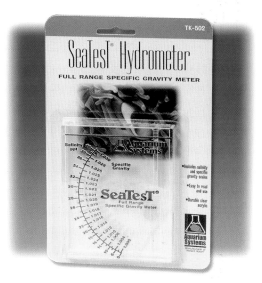

Dip and read hydrometer.

===== TIP =====

Heating Seawater

If an extra aquarium heater is unavailable to warm stored seawater to aquarium temperature, make up your seawater supply using only half the final volume of water, and dissolve the salt mix completely, creating a brine. When you need to add water to the aquarium, heat demineralized water on your range or in your microwave oven. Use the hot water to dilute the brine previously prepared until you have the correct salinity and temperature. Never heat seawater to the boiling point, as this will alter its composition.

usually after a specified number of gallons (liters) have been filtered. Carbon prefilters are supplied as cartridges and fit into a housing identical to that for the mud and sand filter.

Conductivity monitors are available for all types of water treatment systems. These instruments indicate when it is time to replace the RO membrane or the DI resins. One can also periodically test the product water for total hardness with an aquarium test kit. A reading higher than 5 ppm probably means it is time to service the unit. A gauge for monitoring line pressure prevents damage from overpressurization by alerting you that something is amiss and will also indicate when an RO unit needs a new membrane. As the membrane becomes clogged with impurities, back pressure increases. A sudden rise above the normal line pressure should be taken as an indication that

the membrane may need replacement. An obstruction or kink in one of the water lines will also raise the pressure, and this possibility should be investigated to prevent damage.

A variety of containers can be used for storing demineralized water. Plastic household storage containers with lids are adequate for 10 to 20 gallons (40 to 80 L). For larger quantities, why not use a 30- or 50-gallon (120- or 200-L) plastic garbage can with a snap-on lid?

Salt Mixes

If properly treated after collection, natural seawater is the best choice for filling a marine aquarium. Aquarists who live away from the shore may find it impossible to obtain natural seawater, and even those in coastal cities may find the cost or the trouble prohibitive. For aquarium use, seawater should be collected offshore, away from both the diluting effects of estuaries and the pollution supplied by onshore development. After collection, the water must be aged for a week or two to allow planktonic organisms to die and to permit particulate matter to settle. Filtration via a diatomaceous earth (DE) filter, available at most aquarium shops, can speed up the process, as can the introduction of ozone or other sterilants.

One marine fish hatchery located on the coast used chlorine to disinfect its seawater supplies, which were held in a reservoir the size of a small swimming pool. To fill the reservoir, water was pumped in through a pipe located about 500 feet (150 m) out to sea. Pool chlorinator was then added, and the water was allowed to stand for 24 hours. Afterward, the chlorine was neutralized via the addition of sodium thiosulfate (the active ingredient in dechlorinators sold in aquarium shops). Particulate matter was removed by filtration before the water was introduced into the hatchery's aquariums.

In coastal areas, for example, Los Angeles, seawater collectors sell water for public aquariums, restaurants with seafood-holding tanks, and even some aquarium shops. The water is collected at sea by small tankers that filter and ozonate the water before delivery. Such sources may be more costly than preparing synthetic seawater from a dry mix. However, using natural seawater appears to be worth the added cost, judging from the experience of aquarists who do so.

Some of the major and minor components of natural seawater and the values reported for their concentrations are found in Table 4 on page 83. Despite the claims of salt mix manufacturers, natural seawater has never been exactly duplicated via an artificial formulation.

Preparation and Storage

Most aquarists will find that using artificial seawater offers the most practical way to supply the requirements of a home aquarium system. In fact, the introduction of an inexpensive, mass-produced seawater salt mix helped to spur the popularity of marine hobby aquariums in the 1970s. These days, the aquarist has several brands from which to select.

When a packaged dry salt mix is rehydrated, the resulting solution may or may not adequately mimic natural seawater. One reason for this are the differences among brands due to differing formulations. Another is the variation from batch to batch that occurs during the manufacturing process. Still another is the separation of component compounds due to settling during storage and shipment. Avoiding the latter problem merely requires rehydrating the entire container of mix all at once, but little can be done to compensate for the former ones. Prudent aquarists will therefore routinely check freshly made seawater and keep a record of the results. It is worth mentioning that the author has used five brands of synthetic seawater mix over the past 25 years, and all gave completely satisfactory results.

To prepare synthetic seawater, select a suitable container with a cover, and combine the appropriate amount of dry mix with demineralized water. You will need 35 grams per liter, or about 5 ounces per gallon. Manufacturers supply salt mix in premeasured packaging, often labeled as sufficient to prepare 5, 10, 25, or more gallons (20, 40, or 100 L or more). Some calculate this volume, however, based on a final salinity of less than 35 ppt. Therefore, the package will actually produce somewhat less full-strength seawater than advertised. This saves a few pennies per bag in manufacturing costs and increases the producer's profits accordingly. Read labels carefully, therefore.

Although stirring the mixture now and then while allowing it to sit for a day or two will solubilize the seawater mix, one can speed up the process by agitation and applying heat. An air pump and air stone can be used to mix the solution as can a small submersible pump or powerhead. Use a submersible aquarium heater to raise the temperature of the reservoir to correspond to that of the aquarium.

Maintain agitation of the reservoir for at least 24 hours. At that point, test the water for pH, alkalinity, calcium, phosphate, and nitrate. The latter two compounds should be undetectable at the limit of a hobbyist test kit,

Fanged Blenny, Meiacanthus oualanensis.

Queen Angelfish, Holacanthus ciliaris.

Passer Angelfish, Holacanthus passer.

Flagfin Blenny, Acanthemblemaria hancockii.

Speckled Blenny, Cirripectes stigmaticus.

Above left: Lionfish,
Pterois volitans.

Above right: Seahorse,
Hippocampus histrix.

Left: Squirrelfish,
**Sargocentron
xantherythrurus.**

while the other three parameters should be within the ranges specified in Table 1 on page 81. Note that these tests should be performed after adjustments to bring the synthetic seawater to a salinity of 35 ppt.

Seawater keeps indefinitely if stored in a covered container in a cool, dark place. Having enough water on hand to perform a 20 percent water change is advisable.

Salinity

Since the 1960s, a formal definition of salinity, referenced to a standard solution of potassium chloride (KCl) at 15°C and 1 atmosphere of pressure, has enabled scientists throughout the world to compare notes on the chemistry of seawater. A value, K_{15}, is determined according to the following formula:

$$K_{15} = \frac{\text{conductivity of seawater sample}}{\text{conductivity of KCl standard}}$$

The K_{15} value is related to the salinity, S, by an enormously complex algebraic expression that no aquarist needs to understand, or even to use, in making salinity determinations.

Several practical methods to determine salinity are available. The most widespread of these among aquarium hobbyists involves estimating the salinity from a determination of specific gravity. While this approach is both quick and simple, aquarists who fail to comprehend the relationship between specific gravity measurements and salinity can be operating their aquariums at significantly lowered or elevated salinity, with consequent stress for the inhabitants.

Much of the popular aquarium literature incorrectly quotes specific gravity readings, not salinity, when offering recommendations. Too many hobbyists overlook the need for comparing specific gravity readings at the same temperature to obtain equivalent results. Specific gravity determinations yield the ratio of the weight of 1 liter of sample to 1 liter of pure water. Since this ratio is based on a standard volume, temperature must be taken into account because the sample will expand or contract as its temperature varies.

Specific gravity may be measured by means of a hydrometer. Some hydrometers are designed to float higher as specific gravity increases. One reads specific gravity directly from a scale at the point where it intersects the water level. To facilitate this type of measurement, a water sample should be removed from the tank to a tall, narrow vessel into which the hydrometer is inserted. Simply floating the instrument in the tank, the approach usually taken by the inexperienced, not only makes it difficult to read but also invites accidental breakage. Another problem that occurs, especially with cheaper glass hydrometers, renders them inaccurate when the scale, printed on a piece of paper glued inside the tube, becomes detached and shifts up or down. This type of hydrometer can be easily recalibrated, however, by simply taking a reading from a sample of pure water, which should have a specific gravity of 1.0000. Any deviation can be subtracted from or added to a sample reading as a correction factor.

A more popular design is the dip and read hydrometer, consisting of a floating pointer inside a clear plastic box. A scale engraved onto the box reads specific gravity according to where the pointer comes to rest. Manufacturers' claims indicate that the scale is calibrated to the operating temperature of a typical marine aquarium, around 75°F (23.8°C). However, a new instrument should be calibrated against one of known accuracy before accepting the readings. Experience indicates that considerable variation exists from one unit to the next, even between those from the same manufacturer. Deposits accumulating on the pointer, or air bubbles adhering to it when a reading is taken, can alter the reading obtained. Age seems to result in decreased accuracy, perhaps because the pointer pivots less freely with continued use.

Despite the drawbacks of using hydrometers to determine salinity, with careful attention to proper procedure, reasonable results can be obtained. Furthermore, this is clearly the method of choice for the vast majority of hobbyists, probably because a hydrometer is relatively cheap compared with the other instruments used for measuring salinity. Tables 2 and 3 on page 82 may be used to convert any observed hydrometer reading to salinity levels when the temperature is also known. A temperature conversion factor is first added to the last two digits of the observed reading to yield the density of the sample. Salinity may then be read directly from a density to salinity conversion table.

The *refractive index* of a water sample varies with density, thus providing another way to measure salinity. (Refractive index is a measurement of how a transparent substance bends light passing through it. An everyday demonstration can be carried out with a glass half filled with water and a pencil. Place the pencil into the glass, and observe from the side. You will note that the pencil appears to be bent where it enters the water. This is because light travels more slowly in water than in air. As a general rule, substances with a higher density slow down the light more and consequently have a higher refractive index.) An instrument, the refractometer, provides a direct reading of salinity when properly calibrated. The refractometer is an optical device and typically costs considerably more than a hydrometer. After initial calibration with freshwater, numerous salinity determinations may be made with the same instrument before recalibration becomes necessary. One only needs to apply the sample, look through the eyepiece, and read the salinity from the scale. With care, a refractometer can last a lifetime.

Increasingly, electronic conductivity meters are used for salinity measurement, providing a digital readout. Units intended for aquarium use are reasonably priced. Electronic equipment, of course, requires proper maintenance and calibration. In the long-run, though, it can be the most convenient method of obtaining precise measurements quickly.

Regardless of the method of measurement employed, adjustment of synthetic seawater to a salinity of 35 ppt should be carried out prior to using the water or performing chemical tests. By standardizing the salinity of all the water used in the aquarium, one may be assured that test results will be similar from one batch of water to the next and that the aquarium's inhabitants will be exposed to minimal fluctuation in conditions.

BIOCHEMICAL CYCLES AND FLUCTUATING CONDITIONS

Marine aquariums develop into miniature ecosystems that can mimic the ocean to a remarkable degree. In the marine aquarium as well as in the sea, dynamic, cyclical processes affect water chemistry. Waste detoxification and acid-base balance are critical to the survival of captive marine life.

Effects of Microscopic Life

A tank of synthetic seawater becomes an aquarium with the introduction of living organisms. Among the first of these to arrive are bacteria, algae, fungi, and other single-celled organisms that may be wafted in on a current of air. Typically, marine microorganisms are added deliberately. Such creatures are abundantly present in and on the *live sand* and *live rock* taken directly from the sea. They are used to seed the developing aquarium. Even in the relatively sterile systems that are sometimes constructed to display a collection of

In a balanced marine system, reef fishes thrive. (Blue Spotted Jawfish, **Opistognathus rosenblatti**.)

larger reef fish, microscopic life coats every available surface and contributes to the artificial ecosystem that a functioning aquarium becomes.

Although ecologists might find fault with the notion of calling an aquarium an ecosystem, aquariums do resemble genuine ecosystems in many respects. Much of the resemblance can be accounted for by the activities of microscopic life that persists in performing its ecological function whether at home in the sea or confined within a tank. Marine reef aquariums exhibit *trophic levels,* or food webs, in which photosynthesizers provide food for primary consumers that, in turn, are eaten by secondary consumers. They also exhibit *biochemical cycling,* the utilization of one organism's waste products as food for another. These dynamic phenomena are remarkably similar to those associated with natural coral reefs.

Any organism placed within the confines of an aquarium immediately begins to change the chemistry of the water. Respiration, for example, removes dissolved oxygen and adds carbon dioxide. Carbon dioxide tends to disrupt the acid-base balance of the system, although the buffering action of seawater compensates for this effect. Oxygen removal by one organism means less oxygen availability for another organism unless replenishment takes place continuously. Excreted nitrogenous waste threatens to poison most creatures unless it is rendered harmless by bacteria. Monitoring these processes through regular chemical analysis of the water therefore constitutes an important part of routine aquarium maintenance.

To understand better how these cycles work and how they are interlinked, first consider a simple food web that might take place in a marine aquarium. Algae growing on a rock are cropped by a damselfish. The fish extracts oxygen from the surrounding water and uses it to metabolize the carbohydrates and proteins contained within the algae. Besides carbon dioxide, the fish excretes nitrogen in the form of ammonia along with its solid waste, mostly undigested cellulose from the algal cell walls. When the aquarium ecosystem functions properly, each of these potentially toxic waste products is recycled by microorganisms, algae, and invertebrates. An improperly functioning system may signal distress with accumulations of wastes, a drastic change in pH, or other ways that may be detected by water tests.

Nitrogen Cycle

Accumulation of nitrogenous waste poses the most significant threat of immediate harm to aquarium inhabitants. Ammonia, NH_3, is the primary excretory product of marine organisms. It is measured as total ammonia-nitrogen (NH_3-N). Its level in the marine aquarium should not be allowed to rise above about 0.1 mg/L. This is below the level of detection of most aquarium test kits. Fortunately, microorganisms known as nitrifying bacteria can be easily cultivated in the aquarium. These bacteria convert ammonia into the much less toxic nitrite (NO_2^-) and nitrate (NO_3^-).

Biological filtration refers to the detoxification of aquarium water by nitrifying bacteria. In simplified form, the equations for this two-step process are

$$NH_4 + 2O_2 \rightarrow NO_2^- + 2H_2O$$
$$\text{(equation 7)}$$

$$NO_2^- + O_2 + 2H^+ \rightarrow NO_3^- + H_2O$$
$$\text{(equation 8)}$$

Ammonia is excreted as a waste product by the fish and invertebrates that occupy the tank. This ammonia serves as an energy source for a group of nitrifying bacteria that oxidize it to nitrite (equation 7). A second group, utilizing the nitrite as its energy source, oxidizes it to nitrate (equation 8). Without this form of biochemical cycling, marine aquariums would be impossible. Questions remain regarding the exact identity of the microbes responsible for aquarium *nitrification* (a general term for the overall process of oxidizing ammonia to nitrate).

Furthermore, experienced aquarists may disagree about the best method for establishing a suitable population of nitrifying bacteria or regarding which system design best facilitates the nitrification process. However, nearly all aquarists agree about the basic principles by which nitrification makes an aquarium possible. Obtaining the necessary bacteria is no problem. One needs to add only a small amount of live rock, live sand, or any material taken from the ocean or from an aquarium with an established biological filter. Biological filtration is facilitated when there is an abundance of surfaces in contact with oxygenated water available within the aquarium system. The nitrifying bacteria must attach themselves to a solid object (millions of them can be on a grain of sand) in order to carry out their chemical conversions. They must also have plenty of oxygen since nitrification is an aerobic process. This can be readily deduced from equations 7 and 8. (Organisms that require oxygen, such as nitrifying bacteria and human beings, are said to be *aerobic*. The term is also applied to biological processes that demand oxygen.) Apart from these relatively simple requirements, the bacteria will work their chemical magic with no additional encouragement from the aquarist.

This knowledge may be put to practical use in several ways. For example, if one wishes to determine if sufficient biological filtration capacity exists to process a reasonable amount of ammonia from the fish population that will soon be introduced, one can challenge the system with a dose of ammonia in chemical form. By means of an ammonia test kit, an aquarist can follow the ammonia-nitrogen (NH_3-N) reading while introducing a small amount of ammonium chloride solution (NH_4Cl) to raise the level to 3.0 mg/L NH_3-N.

Simultaneously, a nitrate-nitrogen (NO_3-N) test should be performed on another sample of water taken from the aquarium. After several hours, repeated ammonia and nitrate tests taken at regular intervals yield numbers that, when graphed, will demonstrate an increase in nitrate as ammonia decreases. When the ammonia-nitrogen concentration has decreased to zero, the nitrate-nitrogen concentration will have increased by 3.0 mg/L, exactly equivalent to the amount of ammonia-nitrogen added at the beginning.

In actual practice, of course, not all of the ammonia may be converted to nitrate. Some might be taken up by algae living in the aquarium, and the biological filtration process will not achieve the theoretical maximum efficiency. Nevertheless, the end result will approximate what has just been described.

Because nitrification takes place in two steps, accumulation of the intermediate compound, nitrite, can be taken as an indication of disruption. The oxidation of nitrite to nitrate yields less energy for the bacteria than does the oxidation of ammonia. The bacteria responsible succumb first to perturbations in the system. In the event one suspects trouble, a quick check for the presence of ammonia or nitrite should be immediately carried out.

An accumulation of ammonia is extremely serious. If this happens, fish should be immediately relocated to an operable system. Aquarists with a large collection of fish are strongly encouraged to maintain a second tank for this purpose, but they seldom do. Regular partial water changes to keep the ammonia in check until the bacteria have a chance to recover offer the only alternative. One can also introduce additional live rock or live sand, either of

Fish with large appetites place more demands on the filtration system. (Orangethroat Pike Blenny, Chaenopsis alepidota.)

A few hardy species, such as the Northern Toadfish, Opsanus tau, can survive in polluted water.

which will bring along nitrifying bacteria. Of primary concern, however, is locating and eliminating the source of the problem.

Overcrowding—placing a larger bioload into the tank than can be accommodated—overwhelms the bacterial population with more ammonia than can be processed. Severe overcrowding may lead to ammonia accumulation. However, borderline overcrowding more commonly results in a chronically low level of nitrite in the water at times. Increasing the ratio of nitrification capacity to bioload offers

Coral Trout Grouper, **Cephalopholis miniatus.**

Powder Blue Tang, **Acanthurus leucosternon.**

the only practical solution to this problem. One must have a larger tank and/or increased biological filtration. Aquarists have turned to a variety of methods to achieve this.

Nitrifying bacteria must attach themselves to a solid surface and must have an abundant supply of oxygen in order to perform adequately in maintaining the aquarium environment. If the system provides an insufficiency of either attachment sites or dissolved oxygen, nitrification will be impaired. Prior to the mid-1980s, the undergravel filter was employed almost universally. The gravel on the tank bottom served as the filter bed. Water was pumped up from below the gravel bed and discharged near the surface, where it could pick up oxygen, while freshly oxygenated water flowed downward through the bed, to rid it of its ammonia. Unfortunately, debris easily accumulated in the gravel, impeding the flow, reducing filtration efficiency, and actually contributing to the bioload as it slowly decayed. To maintain the system at optimal nitrification capacity required suctioning the debris from the filter bed periodically, a chore relished by no one.

Around 1985, a new filtration system design spurred renewed interest in marine aquarium keeping and created what can only be described as a revolution. The wet/dry filter, also known as the trickle filter, is an innovation borrowed from sewage treatment technology. It avoids the debris accumulation problem by moving the bacterial culture to trays of gravel and later to a chamber containing specially designed plastic pieces, over which tank water is sprayed. As the water drips over the trays, it is both aerated and stripped of ammonia by nitrifying bacteria. The filtration trays and

TIP

Interpreting Test Results

Confusion can occur when interpreting test results for ammonia or ammonium, nitrite, and nitrate. This is because the concentrations of these substances can be expressed in two ways: as the ion or as the amount of nitrogen present. If the test kit measures total ammonia/ammonium (NH_3 or NH_4^+), divide by 1.3 to obtain the ammonia-nitrogen reading (NH_4^+-N). If the measurement is of nitrite ion (NO_2^-), divide by 3.3 to obtain the nitrite-nitrogen (NO_2^--N) level. If the kit records nitrate ion (NO_3^-) levels, divide by 4.4 to obtain the nitrate-nitrogen (NO_3^--N) reading. To convert nitrogen readings to ion equivalents, multiply by the respective conversion factors.

Tangs are intolerant of accumulations of organic matter in the water. (Yellow Tang, Zebrasoma flavescens.)

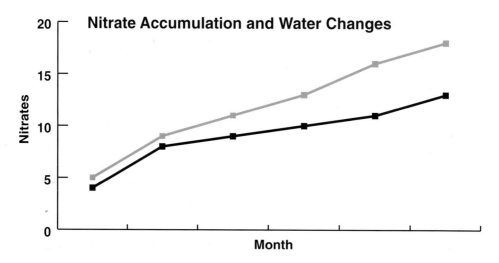

Nitrate Accumulation and Water Changes

Graphic representation of the effect of monthly water changes on the nitrate level in a typical marine aquarium. The data assume that the tank accumulates 5.0 ppm of nitrate each month and that a 20 percent water change is carried out each month. The gray line indicates the nitrate concentration prior to the water change, while the black line indicates the concentration after the water change. Note that nitrate steadily accumulates despite water changes.

chamber are not immersed, hence the term wet/dry filter. Filtered water accumulates in a sump below the wet/dry chamber and from there is pumped back into the aquarium. Because the filter bed is suspended above the sump, debris does not clog it but instead accumulates in the sump, from which it can be easily removed.

One reason for the success of the wet/dry filtration systems is their high degree of nitrification efficiency, enabling larger numbers of fish to be maintained in less space than was the case with undergravel filtration. Another reason is the association of this type of system with a completely new kind of marine aquarium, the minireef. This features not only colorful reef fish but also numerous species of invertebrates and plants. To say that the new hobby of reef keeping struck the aquarium industry like a bolt of lightning is an understatement. Dozens of new businesses sprang up overnight, either to manufacture wet/dry filter systems or to market dozens of other innovations. Despite all the hullabaloo, reef aquarium enthusiasts today have largely abandoned wet/dry filters. They prefer system designs that are more natural in their approach and that clearly represent improvements over the wet/dry filter in maintaining optimal water chemistry.

Several conditions can lead to an interruption of nitrification as opposed merely to overwhelming the system's capacity. However, the symptoms are the same: ammonia, nitrite, or both accumulate, and the tank's inhabitants exhibit signs of stress. The addition of antibiotics in an attempt to cure disease frequently results in damage to the nitrifying bacteria. Antibiotic therapy is seldom effective in the home aquarium because of inadequate diagnosis, insufficient dosage recommendations, the existence of resistant strains, and degradation of the antibiotic in storage. Worse, reduced biological filtration capacity may compound any disease problems. Under no circumstances should any form of medication or therapeutic agent be added to a display aquarium with an actively functioning biological filter. Treatments should always be carried out after transferring affected animals to a separate hospital tank.

Nitrification may also be interrupted due to reduced oxygen levels or temperature extremes. An equipment malfunction or power failure is usually at fault. If water circulation is interrupted, oxygen cannot enter the water from the atmosphere. If the temperature rises too high, the water cannot retain enough oxygen. As the temperature falls, even though the ability of the water to hold oxygen increases, bacterial activity slows and then stops. At refrigeration temperatures, viable nitrifying bacteria may survive for several months although their metabolic activity essentially ceases. Nitrification usually resumes, after a short lag period, once aberrant conditions are corrected. However, transient low levels of nitrite or ammonia may appear in the aquarium during the interruption. Accurately predicting the effects of such temporarily impaired water quality is impossible. In such situations, perform a partial water change if this will help to restore normal conditions.

Observe all the tank's inhabitants closely over the next week in case any symptoms appear.

The end products of nitrification—hydrogen ions and nitrate—affect the aquarium environment differently as they accumulate. Hydrogen ions will be discussed later, in connection with acid-base balance. At this point, let us turn our attention to nitrate.

Nitrate is not considered toxic to many marine organisms. Some fish can tolerate nitrate in concentrations that are thousands of times greater than they would ever encounter in the ocean. However, many authors recommend relatively low levels of nitrate, usually around 20 mg/L NO_3^- or less. Nitrate concentrations near natural reefs are quite low. Measured values for total inorganic nitrogen in tropical seas are typically in the range of 1–2 micrograms per liter. That equals 1–2 parts per billion for the sum of all ammonia, nitrite, and nitrate present. For the aquarist using simple test kits, this number is effectively zero. Irrespective of the effects, if any, on the aquarium's inhabitants, nitrate monitoring offers a convenient way to observe the *nitrogen equilibrium* of the system.

An aquarium may be said to be in nitrogen equilibrium if the amount of nitrate detectable in the water remains constant with time. Nitrogen enters the aquarium as food. That which does not become animal protein in the consumer is excreted as ammonia and is eventually oxidized to nitrate via biological filtration. Nitrogen leaves the aquarium in one of three ways. Regular exchanges of water as part of routine maintenance reduce nitrate concentration via dilution. Unless 100 percent of the water is changed, however, nitrate will steadily accumulate, its concentration forming

Achilles Tang, **Acanthurus achilles.**

a sawtooth pattern trending ever upward if plotted against time on graph paper.

Algae provide a second avenue for nitrate removal. By absorbing nitrogen as ammonia, algae circumvent the biological filtration pathway. When algae are removed from the tank during cleaning, the nitrogen they have absorbed is eliminated from the system.

The third mechanism, known as *denitrification,* is facilitated by another group of bacteria. Denitrifying bacteria perform the opposite of nitrification, reducing nitrate to nitrogen gas, which escapes into the atmosphere. Although oxygen is essential for nitrification, it impedes denitrification because it is poisonous to the bacteria involved. (Organisms that survive in the absence of oxygen are said to be *anaerobic,* as are the biological processes they mediate, such as denitrification.) The process occurs, therefore, either in sediments at the bottom of the tank or deep within the recesses of rocks and coral, where oxygenated water does not circulate. A certain amount of denitrification thus occurs in any aquarium that affords an area of anoxic water.

Furthermore, systems can be designed to maximize denitrification, reducing the level of nitrate below the limit of detection for aquarium test kits. Denitrification can thus be the major avenue for nitrate removal from an established aquarium system. One may speak of the *denitrification potential* of the aquarium as the capacity of anaerobic bacteria present in the system to reduce nitrate. When the denitrification potential equals the production of nitrate by the nitrification process, the amount

Yellowtail Squirrelfish, **Neoniphon aurolineatus.**

Clown Wrasse, **Coris gaimard.**

Clown Anemonefish, **Amphiprion percula.**

of nitrate measured in the water will remain constant. The system may then be regarded as being in nitrogen equilibrium. Any perturbation of either nitrification or denitrification will, therefore, be recognizable as an abrupt change in nitrate concentration.

A transient increase might occur, for example, if a new fish is introduced to the tank. After a lag period, the bacterial populations will adjust to the new bioload, and equilibrium will be restored. On the other hand, if nitrate levels rise

month after month despite partial water changes, not enough denitrification is occurring. The aquarist might then consider adding live rock to provide more denitrification potential.

Nitrate digesters are another option. These devices are designed to facilitate denitrification in a manner analogous to the enhancement of nitrification via wet/dry filters. They are, unfortunately, rather costly and require frequent, careful maintenance to operate properly. Most hobbyists opt to facilitate both nitrification and denitrifi-

Pygmy Angelfish, **Centropyge argi.**

Flameback Hawkfish, **Cirrhites acanthops.**

cation with the more natural means of live rock and live sand and to maintain efficiency by regular, consistent maintenance procedures. These include removal of debris, partial water changes, and adjustment of chemical parameters to maintain normal ranges. Not only does the natural approach maintain a healthy environment for the aquarium's inhabitants, it also costs less and reduces maintenance time when compared with technologically elaborate alternatives.

Carbon Cycle

Most people are probably familiar with the carbon cycle from high school biology. It is the fundamental way in which the Sun's energy fuels Earth's living systems. In brief, photosynthetic organisms capture sunlight and use it to convert molecules of carbon dioxide and water into sugars. Nonphotosynthetic organisms are unable to create food from sunlight and simple molecules. They must therefore consume photosynthetic organisms, or each other, in order to obtain energy and building materials for their own metabolic processes. When consumer organisms metabolize sugar by utilizing atmospheric oxygen, excess carbon is expelled into the environment as carbon dioxide and becomes once again available for photosynthesis. This great cycle has kept most of Earth's ecosystems operating for hundreds of millions of years.

In an aquarium setting, the situation is somewhat different. Far more carbon enters the system as food deliberately introduced by the aquarist than is trapped via photosynthesis taking place in the tank. It is certainly true that algae growing in the aquarium trap carbon dioxide respired by the animals and that when these algae are subsequently consumed,

the carbon is recycled, mimicking the natural process. As a general rule, however, aquarium systems do not achieve the capacity to recycle all of the added food or to produce internally enough food to meet the needs of the animals. An excess of carbon compounds dissolved in the water results from this imbalance, requiring offsetting action on the part of the aquarist.

DOC: The collective term for the various carbon compounds accumulating in aquarium water is *dissolved organic carbon* (DOC). Aquarists seldom test the concentration of DOC, as there is no convenient, quantitative way to do this at home. Instead, they simply design the system to minimize the accumulation of DOC and perform maintenance to reduce its concentration. The concern here is the increased bioload that results from decomposition of DOC. As DOC slowly decomposes, ammonia is released. This ammonia must be processed by the biological filter, depleting oxygen and adding hydrogen ions and nitrate to the water. In effect, having a large amount of DOC present is like adding additional fish to the tank.

Systems with large amounts of natural microbial activity appear to generate less DOC, probably because the diverse population of microorganisms accomplishes carbon recycling efficiently. Care in adding food to the system will obviously have a negative effect on DOC. Choosing foods that are fresh, appetizing, and appropriate to the species being fed will also negatively affect DOC. All of these help ensure that food will actually be consumed, rather than falling to the bottom to decay slowly. Even a combination of all these approaches will not reduce DOC to natural reef levels, however. To accomplish this, many aquarists utilize a piece of equipment called a *protein skimmer*.

Protein skimmers: If you have ever seen billows of foam blowing across the sand at the seashore, you have observed the operation of the ocean's natural protein skimmer. Although this device has been around for decades, only since the 1990s has its value in maintaining minireef aquariums been fully appreciated. The principle upon which protein skimmers operate is quite simple. If fine air bubbles are introduced into seawater containing DOC, the organic molecules tend to collect on the surfaces of the bubbles, forming a dense foam. The process is precisely analogous to the formation of a meringue when a chef introduces a large amount of air by beating a bowl of egg whites with a whisk. The meringue, or foam in this case, becomes more viscous and rises to the surface of the container. In a protein skimmer, fine air bubbles are injected into a column of water pumped from the tank. The resulting foam, which contains the DOC, rises to the top of the column. The skimmer's design permits the foam to overflow into a reservoir, where it accumulates as a greenish brown viscous liquid. The reservoir is periodically emptied, eliminating the DOC.

A major advantage of protein skimmers over other means of DOC removal, such as passing the water over activated carbon, lies in the fact that the DOC is sequestered out of continued contact with the aquarium water. Its decay therefore does not compete with the tank's inhabitants for oxygen. The resulting ammonia is not released back into the system.

Skimmers have achieved such popularity that a bewildering host of designs and brands now exists. Despite advertising hype, no one design appears to be best. In choosing a skimmer, follow the recommendations of your local dealer or a knowledgeable aquarist friend. When in doubt, choose a skimmer designed for a system a little larger than the one you have rather than selecting a smaller one. Overskimming the aquarium is not possible. However, a skimmer that is too small will, at worst, be inefficient and, at best, will require more frequent emptying than one appropriately sized.

Acid–Base Balance

As may be readily appreciated from the description of the primary biochemical cycles taking place, the chemistry of aquarium water always remains in a dynamic state. This is one reason periodic testing is essential. Changes are inevitable. The aquarist needs to know both the rate and magnitude of changes in order to take corrective action. Dynamic processes, including photosynthesis and respiration, and consequently the nitrogen and carbon cycles, affect the acid-base balance of the aquarium. These, in turn, affect the metabolism of each of the tank's living residents. Therefore, routine analysis of pH and alkalinity are essential to successful long-term marine aquarium care.

The pH of a solution is defined as the negative logarithm of the hydrogen ion concentration, written as:

$$- \log [H^+]$$

(equation 9)

To understand the meaning of this equation better, return to the idea of water molecules dissociating into hydrogen and hydroxyl ions:

$$H_2O \rightleftharpoons H^+ + OH^-$$

(equation 10)

In a sample of pure water, only a small proportion, 1/10,000,000 or 10^{-7}, of the molecules dissociate, resulting in a hydrogen ion concentration of 10^{-7} moles/L. The logarithm of a number is the exponent to which the number 10 is raised to equal that number. Thus, the logarithm of 10^{-7} is 7. Substituting this value into equation 9 above yields $-(-7)$, or a pH of 7.0 for pure water. This value of 7.0 is considered neutral, neither acidic nor basic. As acid is added to the system, for example by dissolving HCl in the water, the concentration of hydrogen ions (expressed as $[H^+]$) increases. The pH therefore decreases, falling below 7.0. The resulting solution is said to be acidic. Conversely, a pH higher than 7.0 results from fewer hydrogen ions in solution, and the solution is regarded as basic. Seawater is basic, with a pH of 8.2 to 8.3.

Seawater is also a buffer, meaning it resists changes in pH despite the introduction of acids from respiration and biological filtration. This occurs because ions, such as carbonates and borates, neutralize acid as the acid is added. When a seawater sample is titrated with acid, one can see carbon dioxide escaping as gas bubbles as the acid reacts with carbonate.

Eventually, however, the buffering capacity will be exhausted, and the pH will begin to change. The amount of acid required to titrate seawater to this end point is about 2.5 milliequivalents per liter (meq/L), the value for seawater's alkalinity.

Alkalinity is measured by determining how many equivalents of acid must be added in order to combine completely with the bases present in the seawater sample, without adding an excess. To do this, an aquarist adds a standardized acid solution drop by drop and notes when the end point is reached by means of an indicator that changes color in response to different pH levels. The end point of the alkalinity titration is reached at a pH of 4.5. The indicator dye commonly used in aquarium alkalinity test kits changes from blue to yellow at the end point. Determining the end point can be done more precisely using a pH meter. If one does this, records the solution's pH as each drop of acid is added, and then plots the results, a stair step graph results. The pH remains stable for the first few additions of acid and then drops precipitously as the end point is approached.

Maintaining the alkalinity of the aquarium at the natural seawater level or higher means that the pH will remain stable despite the introductions of acid from biological filtration and the respiration of animals and plants. Hydrogen ions are produced as a direct consequence of the oxidation of ammonia. Respired CO_2 reacts with water to form carbonic acid, which quickly dissociates to produce bicarbonate and hydrogen ions, according to the following overall formula:

Powder Brown Tang, **Acanthurus japonicus.**

$$CO_2 + H_2O \rightleftharpoons H_2CO_3 \rightleftharpoons HCO_3^- + H^+$$
(equation 11)

Partial water changes and the addition of compounds including calcium hydroxide solution and various commercial products may be used to restore aquarium alkalinity.

If respired CO_2 is removed almost as quickly as it is produced, the tank will remain in equilibrium, and there will be no net change in pH. On the other hand, if carbonate and bicarbonate ions are being removed from the aquarium by photosynthesis, the pH may rise as equation 11 shifts to the left. In aquariums with heavy algae growth and bright illumination, the pH can reach 9.0, well outside the range of comfort for many organisms.

Flame Angelfish, Centropyge loriculus.

The effects of photosynthesis on pH deserve special attention because photosynthesis is a light-dependent process. In the dark, plants and algae respire more or less identically to animals and fungi. For this reason, the pH of an aquarium is likely to be at its lowest point immediately before the lights are turned on in the morning and at its highest just as they are being extinguished in the evening. The importance of this fact lies in its implications for test procedure. If pH tests are carried out at different times of day, making sense of any variation becomes impossible. Thus, the aquarist should carry out pH tests at a consistent time in order to compare the results accurately.

Minimizing pH fluctuations: Certain types of system designs minimize pH fluctuations. For example, the system designed by Dr. Walter Adey of the Smithsonian Institution uses an algal turf scrubber to process ammonia biologically. Algae growing on the scrubber absorb ammonia and other pollutants from the aquarium because they grow on special plastic grids exposed to intense artificial illumination. The scrubber is typically illuminated while the

lights over the aquarium itself are off and vice versa. Thus, photosynthesis occurs at a continuous rate. The pH of Dr. Adey's system remains stable both day and night.

Another way to achieve pH stability involves an electronic monitoring and control device. By coupling a pH meter to an electric valve that controls the introduction of carbon dioxide gas into the aquarium, the pH can be continuously maintained within a narrow range. The control unit uses digital circuitry to adjust pH in the same manner that a thermostat controls temperature. If the pH rises too high, the valve opens and CO_2 is pumped into the water. When the desired pH is reached, the valve closes. Unfortunately, a second unit (or a single unit with two monitor/control channels) may be needed to keep the pH from going too low, by pumping in calcium hydroxide solution, $Ca(OH)_2$, instead of carbon dioxide gas. A significant problem with pH control systems is the possibility of the unit failing with a valve in the open position, allowing either CO_2 or $Ca(OH)_2$ to flow into the aquarium continuously, with disastrous results. Fortunately, such accidents rarely occur. Fail-safe options include installing a separate monitor that sounds an alarm when the pH rises or falls outside a pre-set range. Some systems can even dial a phone number and deliver a warning message. Having both high and low pH control channels also offers a measure of fail-safe protection. One control can offset a failure in the other, and the likelihood of simultaneous failure is small.

Most hobbyists will purchase a simple color change–based pH kit for testing their marine tank. Make sure the brand you select is intended for saltwater use because unreliable results will be obtained otherwise. Reliable digital pH meters are now available at prices hobbyists can afford and are a vast improvement over chemical pH kits. When considering one of these units, remember that being able to calibrate easily the instrument against standard buffer solutions of known pH is essential. Purchase a small amount of these *calibration buffers* (pH 7.0 and 10.0) along with the pH meter. Carefully read and

TIP

Expressing Alkalinity

Many aquarium references refer to alkalinity as a *KH value* expressed in degrees (dKH). To convert alkalinity in milliequivalents per liter to dKH, multiply by 2.8. For natural seawater, therefore, the KH value is 7 dKH (2.5 meq/L × 2.8). Marine aquariums should be maintained at an alkalinity somewhat higher than that of natural seawater, about 4 meq/L or 10 dKH. Higher alkalinity offsets the accumulation of acids in the closed system of the aquarium.

follow the manufacturer's instructions to calibrate and use the instrument.

Aquarists rarely think about carbon dioxide accumulating in the water, but it is often a source of problems. One of the basic causes is low output from pumps or power heads. This can occur because too small a unit was chosen or because the pump has slowed due to age or lack of maintenance. Reduced water movement results in poor gas exchange, and carbon dioxide cannot escape the tank to the atmosphere. Virtually all of the gas exchange between the aquarium water and the air above it occurs at the surface. Injecting air with an air stone, for example, will increase water movement due to bubbles disturbing the surface. However, this does not directly introduce much oxygen or remove any carbon dioxide. The primary goal of creating water movement, by whatever means, is to bring water from the lower levels of the aquarium up to the surface where it can come into contact with the atmosphere.

Check the pH and alkalinity of your aquarium regularly. If either remains chronically below normal, determine if carbon dioxide is accumulating. This can be done easily. Remove a 0.5-gallon (1.8-L) sample of water from the tank and aerate it vigorously overnight. After 24 hours of aeration, check the pH of the sample, and check the pH of the tank at the same time. Compare the two readings. If the pH of the tank is more than 0.2 pH units below that of the aerated sample, then carbon dioxide is accumulating. Remedial action is relatively straightforward. Increase water movement by adding air stones, or place one or more small submersible pumps, known as power heads, into the tank. Power heads will provide more turbulence if they are arranged so their out-

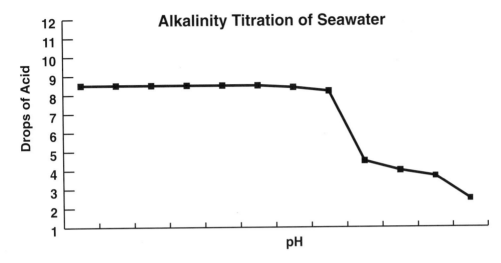

Graphic representation of the titration of seawater with hydrochloric acid. Note that the pH remains stable for the first several additions, then drops suddenly, and delinces steadily thereafter. Such a pattern is typical for a solution that acts as a buffer.

puts intersect rather than having both pointed in the same direction. Increasing the size of the main system pump, if one is used, will also increase water movement proportionately. If these efforts fail, too many organisms are probably respiring within the aquarium. The only remedy in that case is to reduce the population by moving some specimens to another tank.

Summary

By mimicking natural processes, the artificial ecosystem of the marine aquarium develops cycles of oxidation and reduction of nitrogen compounds and carbon. The acid-base system operates in a manner similar to that of the ocean. Understanding these biochemical cycles and their implications in the closed system of a home aquarium is essential to developing a maintenance routine that achieves optimal water quality. When these interacting, dynamic processes are out of balance, the effects on the aquarium resemble those that might be observed on an isolated portion of the sea where human activity has disrupted the ecosystem. High levels of nitrate, phosphate, and DOC are associated with algal blooms, dead corals, and low animal diversity, both in the ocean and in the aquarium.

DISSOLVED COMPONENTS

Substances dissolved in seawater can have profound biological effects. Oxygen, calcium, phosphorus, iodine, and trace ions each affect marine organisms, depending upon their concentrations. Relatively small fluctuations in the concentrations of dissolved components can have large consequences for captive marine life.

Oxygen

Every living organism requires oxygen to sustain life. Oxygen deprivation can create significant stress. As mentioned in the previous chapter, oxygen and other gases enter or leave the aquarium water at the surface. Any marine tank must have good water circulation that brings water up from the lower levels to be exposed to the atmosphere. Aeration with air stones, or more commonly, creating turbulence with one or more water pumps, is necessary to achieve a sufficient oxygen concentration. Seawater at 75°F (23.8°C) contains about 6.8 mg/L of dissolved oxygen at 100 percent saturation (meaning that the water holds all the oxygen theoretically possible at that temperature). An aquarium should have sufficient aeration to achieve this concentration or higher, as levels

Both fish and invertebrates need abundant oxygen. (Longnosed Hawkfish, Oxycirrhites typus, and unidentified red gorgonian coral.)

of *supersaturation* can occur in the reef environment. (When supersaturated, the sea actually contains more oxygen than is theoretically possible. This is due both to vigorous turbulence and to a high rate of photosynthesis, which liberates molecular oxygen directly into the surrounding water.)

Check the dissolved oxygen (DO) concentration when the tank is initially set up and each time new specimens are added. Wait about 24 hours after adding new animals before checking the DO. Wet chemical DO tests require careful attention to the collection of a sample and proper use of several reagents in order to produce accurate results. Electronic DO meters are much simpler to use. The reading obtained with a meter is also likely to be more accurate than what can be obtained by means of a chemical test. Many hobbyists choose not to measure DO routinely because of the considerable expense involved. If water circulation is vigorous, there is probably sufficient DO.

Besides insufficient water movement, other causes of low DO are excessive heat (> 80°F/26.7°C), decomposition of food or a dead organism (decay bacteria consume the available oxygen), and overcrowding.

Calcium

Corals, crustaceans, mollusks, and calcareous algae all extract calcium from seawater to construct their skeletons of calcium carbonate. Calcium is also essential to vital biological processes in the living cells of every organism. Minireef hobbyists have long been aware of the need to replenish calcium periodically, to compensate for its removal both by growing organisms and by chemical precipitation.

Before marine aquarium chemistry was better understood, hobbyists often focused too narrowly on maintaining the level of calcium alone and ignored the role of alkalinity. However, the availability of carbonate, the other essential component in skeletal structures, mostly depends upon both pH and alkalinity. When these two parameters are within the appropriate range, skeleton building can still occur even when calcium is present at a level significantly below that of natural seawater. On the other hand, when *both* alkalinity and calcium concentrations are low, corals do not thrive.

Too much calcium may be worse than too little. Increasing the calcium level above about 550 mg/L will result in precipitation of calcium carbonate and an abrupt decline in alkalinity. Ironically, with lowered alkalinity, calcification is made more difficult, just the opposite of what calcium supplementation should promote.

The *calcium hardness* value for natural seawater is about 400 mg/L. Maintaining this level is a relatively simple matter, as any soluble compound containing calcium can be added to the water to compensate for a deficit. Calcium chloride, calcium hydroxide, and a host of other compounds and mixtures of compounds have been recommended in the aquarium hobby literature. Numerous commercial calcium supplements are available. When choosing, one must remember that ions always come in pairs. Therefore, any compound containing calcium ions also contains partner ions. The nature of these partners can have important implications for aquarium chemistry. For example, using calcium chloride may result in the need for another additive to restore alkalinity, because it is easy to overdose. As mentioned above, this can have unintended negative consequences. Many hobbyists have found limewater, a saturated solution of calcium hydroxide, $Ca(OH)_2$, to be the most convenient way to replenish calcium.

Besides aiding in the maintenance of pH, alkalinity, and calcium concentration, addition of limewater to the minireef aquarium has other benefits. One is the nearly complete precipitation of phosphate (PO_4^{3-}). Phosphate interferes with calcification because it is a crystal poison. If phosphate is incorporated into the developing skeletal lattice, formation of calcium carbonate crystals is inhibited. Phosphate is detrimental to corals under natural conditions if present in excess. Furthermore, undesirable algae growth is almost always a consequence of excess phosphate.

As was pointed out earlier, nitric acid (HNO_3) production is characteristic of biological filtration. Acid production depletes both calcium and alkalinity from the aquarium water via the neutralization reaction and ion-pair formation.

As a result, systems that accumulate nitrate will also experience problems with maintaining pH, alkalinity, and calcium hardness. The tendency to accumulate nitrate can be overcome by carrying out frequent partial water changes or by denitrifying the aquarium.

Denitrification is carried out by beneficial bacteria in several genera, including *Bacillus*, *Denitrobacillus*, *Micrococcus*, *Pseudomonas*, and *Thiobacillus*. These bacteria must have *anaerobic* (oxygen-free) conditions in order to carry out their chemical conversions. They are able to extract energy from the nitrate molecule, converting it to other chemical forms. Most importantly for the aquarist's purposes, they convert it into nitrogen gas that ultimately escapes from the aquarium into the atmosphere. The overall reaction for this process is

$$2NO_3^- + 4H^+ + 2CH_2O \rightarrow N_2 + 2H_2CO_3 + 2H_2O$$
$$\text{(equation 12)}$$

Note that four moles of hydrogen ions are used up for every mole of nitrate ions converted. Thus, the denitrification process also helps to retain alkalinity by using up hydrogen ions (which, if available, would increase acidity). In addition, denitrifiers consume dissolved organic carbon to obtain energy. This process releases carbonates and bicarbonates, further enhancing alkalinity and indirectly promoting calcification.

The ideal pH for the aquarium is 8.4–8.45. The ideal alkalinity is 2.5–3.5 meq/L. The ideal calcium concentration is 412 mg/L. From day to day, and even from hour to hour, these numbers will fluctuate, owing to the dynamic chemical and biological processes occurring in the aquarium continuously. Since the aquar-

---

TIP

Coping with a Large Error Margin
When measuring calcium using a standard titration test, one must allow for a fairly large experimental error. If, for example, the smallest increment that can be discerned (one drop of titrant) represents 20 ppm of calcium, the actual reading is ± 20 ppm. One way to even out such errors is to perform the test in triplicate and average the results.

---

ium, unlike the ocean, is a closed system, the aquarist must intervene to offset the cumulative effects of these changes. The critical parameter for calcification by corals is a stable, high pH, with the optimum being about 8.4. Invertebrates expend energy to pump calcium actively from the water and can make do with levels of calcium that are lower than normal if pH (8.4) and alkalinity (> 2.5 meq/L) are correct. Once an equilibrium is reached, adding calcium ions does not do much because excess calcium combines with carbonate and restores the previous level.

Phosphorus
Phosphorus is like carbon in that it is usually present in great excess in the aquarium, compared with its concentration in the ocean. The phosphate ion (PO_4^{3-}), also known as *orthophosphate*, is the form of phosphorus measured by aquarium test kits. All references to phosphate in this book should be interpreted to

Top left: Marbled Starfish,
Fromia monilis.

Top right: Red Starfish,
Fromia elegans.

Middle left: Tubeworms,
Protula magnifica.

Middle right: Arrow Crab,
Stenorhynchus seticornis.

Left: Royal Gramma,
Gramma loreto.

Cup Coral, **Goniopora** *sp.*

Bubble Coral, **Plerogyra sinuosa.**

Black-Headed Filefish, **Pervagor melanocephalus.**

Gilded Triggerfish, **Xanichthys auromarginatus.**

Blue Boxfish, **Ostracion meleagris meleagris.**

========= TIP =========

Choose the Correct Phosphate Test

Make certain that the phosphate test selected uses the ascorbic acid method. This analytic procedure is the only one that works well in seawater. Some commercially available phosphate tests are intended for use in freshwater only and use an alternate procedure. Such tests will give erroneous results in seawater.

mean orthophosphate concentrations. Particulate organic phosphate (POP), partially degraded organic matter with a high phosphorus content, will also be present and can serve as a reservoir. POP releases orthophosphate ions as it decomposes. Synthetic seawater mix, activated carbon, water conditioners, and food may all be sources of phosphate. Another major source is tap water.

Phosphate is not considered harmful to marine organisms at the concentrations likely to be achieved in the aquarium under normal circumstances. However, this element often lies behind major problems with algae because it acts as a fertilizer. In the ocean, the scarcity of dissolved phosphorus in a form that can be utilized by living organisms tends to be a factor limiting algae growth. Herbivores crop algae from the reef almost continuously, too. In the aquarium, where phosphate is abundant and herbivores are usually fewer in number than they would be in a comparable area of natural habitat, too much algae may grow.

Excess algae may pose a threat to sessile invertebrates because they can smother them or even poison them by exuding toxic substances. Limiting phosphate seems to be an effective way to avoid this unsightly problem.

Ideally, when testing a coral reef aquarium with a kit that is accurate to 0.05 mg/L, the phosphate level should never be detectable. If such a test shows any measurable level of phosphate, one should start looking for ways to reduce the phosphate concentration. A phosphate test kit may be one of the most expensive kits an aquarist purchases.

Iodine

The hobbyist literature contains numerous references to the benefits of adding iodine to the marine aquarium. Iodine in seawater is in its ionic form, iodide (I^-), and is found in natural seawater at a concentration of 0.06 mg/L. It is removed by protein skimming and by activated carbon filtration. It is essential to many invertebrates and fish. For example, certain soft corals that exhibit rhythmic pulsing movements may fall motionless if they lack iodine. Fish, particularly sharks, jawfish, and hawkfish, will develop goiters (an abnormality of the thyroid) if they receive insufficient iodine. Test kits are available in aquarium shops along with additives, such as Lugol's solution, for replenishing iodine. Exercise caution not to overdose.

Trace Ions

An important distinction between the major elements of seawater and the minor and trace elements is that while the ratios of the concentrations of the major elements are constant and therefore unaffected by local conditions, the concentrations of minor and trace elements *are* affected by chemical or biological processes. This is why adding certain ions to, or removing them from, the aquarium has definite biological effects. Examples of minor ions that produce a demonstrable effect when added to the tank are iodide and phosphate. Iodide supplementation can result in accelerated growth of corals and coralline algae, for example. Adding phosphate ions to the aquarium will typically result in an algae bloom.

An example of a trace ion with significant, concentration-dependent effects upon biological processes is copper. Any amount of copper detectable with a hobbyist test kit is lethal to most invertebrates. Yet, copper is an essential element for many life-forms; however, the quantity needed is extraordinarily minute.

Copper is also important to the marine aquarist because it is used as a medication for the most commonly encountered parasitic diseases of marine fish, *Amyloodinium*, known as coral fish disease, and *Cryptocaryon*, known as white spot or marine ick. At a concentration of 0.2 ppm, copper is therapeutic because it kills the parasites. At greater than 0.3 ppm, however, it is toxic to fish. Copper must never be used in an aquarium containing invertebrates and should always be administered in a separate treatment tank. Even if invertebrates are

TIP

Avoid Phosphate Build-up

Purchase a supply of disposable plastic vials in which to carry out phosphate tests. Use a fresh vial for each test. Phosphates are difficult to remove from glass and plastic. They can build up and result in spurious test results. To avoid this, use a new vial each time a phosphate test is performed.

not present, copper should not be used in a display aquarium because the corals, coral rock, and other lime-containing materials present will absorb the copper medication from the water. This will produce a low reading on the test kit, and the unsuspecting hobbyist will add more medication to restore the therapeutic level. Should the pH of the water subsequently fall for some reason, however, the copper will redissolve, dramatically increasing its concentration in the water, with concomitant toxic effects on the fish. Copper intoxication will cause many marine fish to dash madly about the tank, as if trying to escape. Removing the fish to copper-free water will immediately correct the problem.

Good test kits are essential for maintaining correct water chemistry.

Below left: Peppermint Shrimp, Lysmata wurdemanni.

Below right: Anemone Crab, Neopterolisthes oshimanii.

Top left: A watchman goby,
Cryptocentrus sp.

Top right: Spine-Cheeked Anemonefish,
Premnas biaculeatus.

Middle: Yellow Tang,
Zebrasoma flavescens.

Right: Redtailed Wrasse,
Anampses chrysocephalus.

Marine aquariums typically require periodic replenishment of calcium. Calcium may be lost from the water as a result of chemical reactions leading to precipitation of insoluble calcium carbonate. If the marine tank is a miniature reef housing stony corals, giant clams, and other, similar invertebrates, these organisms will constantly extract calcium from the water. Beginning aquarists sometimes assume that coral skeletons, aragonite, and/or calcareous rocks or sand will slowly dissolve, replenishing the calcium content of the water. However, this does not occur

Adding a rounded teaspoon of calcium oxide to a gallon of water produces a saturated solution of limewater.

in most systems. Accordingly, a calcium supplement must be added. Various supplement preparations are available commercially.

Regardless of which one the aquarist selects, monitoring the calcium concentration of the aquarium on a regular basis is important. The aquarium should be tested prior to the first use of any supplement, to establish a baseline reading. Once supplementation begins, periodic testing, say once a week, should be done to ensure that the efforts are achieving the desired result. Most hobbyists aim for 400 ppm of calcium. Published values for natural seawater range from 380 ppm to 412 ppm.

One of the most convenient supplements is limewater, a saturated solution of calcium hydroxide. *Saturated* in this context simply means that as much lime as room temperature water will dissolve completely has been added. It is prepared by mixing lime (calcium oxide) with water. Lime is manufactured in packages intended specifically for aquarium use and sold in pet shops. If stored tightly covered, the dry, white powder keeps indefinitely. Thus, a quantity of limewater may be

conveniently prepared any time it is needed for supplementing the marine tank.

To make 1 gallon (3.8 L) of limewater, place a rounded teaspoon (1.8 g) of calcium oxide into a clean, clear glass or plastic container. Fill the container almost to the top with distilled or RO water. Cap the container, and shake well. Some undissolved powder should be on the bottom of the container after it settles. If not, add a little more calcium oxide, and shake again. In this way, one can be certain that the water has dissolved as much of the lime as possible. Caution! Calcium oxide is somewhat caustic. Do not allow the dry powder to contact your skin, and keep it out of reach of children.

Using limewater for calcium supplementation also helps to maintain the pH and alkalinity of the aquarium because hydroxyl ions from the limewater neutralize some of the acids generated within the system. In effect, this prevents the alkalinity from being depleted, and the pH therefore remains stable. The ideal pH for calcification is 8.40 to 8.45. Allowing the pH to go higher than this will, ironically, impede calcium maintenance, as calcium carbonate will spontaneously precipitate as the pH rises. The result is that the system is robbed of both calcium and carbonate ions. Maintaining the proper balance, that is, adding a sufficient quantity of limewater without overdosing, is best accomplished through the use of an automated system for introducing the limewater and an electronic pH meter to monitor the resulting effect on the acid-base balance of the aquarium. Many aquarists simply use limewater to replace all water that evaporates from the aquarium system. This approach can result in adding about 1 percent of the system volume every day or two but does not appear to create problems.

Some precautions should be observed when using limewater. Old limewater can cause problems, for example. As the solution stands, it accumulates a layer of calcium carbonate on its surface, owing to reaction with carbon dioxide from the atmosphere. Adding the calcium carbonate to the aquarium along with the limewater is a mistake, because the particles of calcium carbonate act as initiation sites for spontaneous crystallization of more calcium carbonate, causing both alkalinity and calcium to drop. This is just the opposite of the desired result of adding the limewater in the first place. Limewater should therefore be prepared in small batches. If the calcium carbonate skin forms, the liquid should be discarded and a new batch prepared. In addition, any accumulation of calcium carbonate observed adhering to the storage container should be completely removed. To do so, pour 1 pint (475 mL) of water into the container and add 2 tablespoons (30 mL) of white vinegar. Let stand for 30 minutes, and the calcium carbonate deposits should be easy to remove with a brush. Make

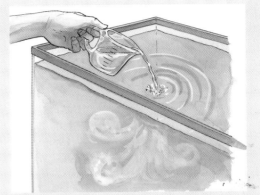

Adding limewater to the aquarium to replace evaporated water is a simple way to replenish calcium ions.

certain to rinse the container well before preparing another batch of limewater. This avoids adding any vinegar, which is acidic, to the system and thus lowering the pH or reducing the alkalinity.

Using impure chemicals for making limewater can also lead to unintended results. Some grades of calcium oxide, CaO, may contain excessive amounts of magnesium, which, when added along with the limewater by the unsuspecting aquarist, results in calcium carbonate precipitation. As mentioned above, precipitation of calcium carbonate removes both calcium ions and carbonate ions from the solution. This therefore lowers not only the calcium concentration but also the alkalinity. However, since magnesium is a component of the seawater buffer system, this situation will also be characterized by a very high alkalinity and high pH, at least initially. Since it is unlikely the home aquarist will be able to test his or her supply of calcium oxide for magnesium contamination, the best approach may be to purchase only products made specifically for aquarium use.

WATER ANALYSIS

Proper marine aquarium management requires periodic testing. Selecting good test kits and using them properly is therefore an important skill for the marine enthusiast. Besides chemical tests, aquarists can also rely upon electronic instruments for fast, accurate assessments of water chemistry.

How Often?

The basic approach to aquarium water testing is similar to the approach a physician uses to evaluate a patient's health. An individual visits the doctor twice a year for a checkup. The doctor makes observations about the condition of the patient's body, weight, blood pressure, and so forth. Knowing what is normal for the individual and observing any significant deviation from those normal values alerts the physician when something is wrong. Exactly the same approach must be taken by the tank doctor, although the checkups should be weekly rather than semiannually.

Keeping track of changing conditions in the marine tank means that the aquarist must periodically carry out water tests and keep a record of the results. In order for the information thus obtained to be of value in maintaining the aquarium, the test equipment used must be sufficiently accurate for the purpose, and the actual testing must be done correctly. Shoddy equipment and/or sloppy technique will inevitably produce results that do not reflect actual conditions in the tank. Such a situation is potentially more dangerous than ignoring tests altogether! Computer programmers have a saying, "Garbage in, garbage out." This means that bad data leads to bad results, even when the computer itself is operating properly. Precisely the same is true for aquarium water testing and its implications for the well-being of the tank's inhabitants.

Choosing Test Kits

For most purposes, a highly reliable test is not absolutely necessary. For example, in the case of ammonia, one needs to know only if it is present at all, not the precise amount, since any amount of this toxic compound is cause for concern. On the other hand, for some tests,

Periodic water testing ensures conditions remain stable for this Dusky Anemonefish, Amphiprion melanopus.

such as copper, a difference of 0.1 ppm can be significant indeed, the difference between treating disease and killing the patient. It is important to understand that what is meant by accuracy depends upon which parameter one is trying to measure.

Accuracy, the degree to which a test result reflects the real state of the sample being tested, becomes ever more dear as the number of decimal places increases. An instrument or test that is accurate to ± 10 units may cost only a fraction of the cost of a similar instrument that is accurate to ± 0.1 units, representing a 100-fold increase in accuracy. Choose a level of accuracy appropriate to the parameter. A calcium test, for example, can vary by ± 20 ppm without causing problems.

Precision refers to the smallest amount of difference a test or instrument will permit the observer to distinguish reliably. The smaller the increment, the greater the precision and the higher the cost. For example, to measure the difference between 0.01 and 0.02 mg/L of phosphate is more difficult than determining the difference between 12 and 24 mg/L. However, since tiny amounts of phosphate can exert dramatic effects, one needs the more precise test.

Which brand should be used? Simple, easy-to-use test kits are sold at any aquarium shop. As with any product, though, not all brands are created equal. Some kits provide results that are almost suitable for serious laboratory work, while others are so inaccurate as to be practically useless. To illustrate the extent to which this is true, the author carried out a simple experiment about ten years ago.

After removing a sample of water from his marine tank, the author tested it with three different test kits. Each was freshly purchased from the local aquarium shop, and each represented a major brand. After performing the tests, the author had the same water sample analyzed by a professional laboratory. The results were surprising. None of the aquarium shop tests were as accurate as the lab's (no surprise there), but only one was close enough to the correct reading to be of any real use. The worst kit gave a reading nearly eight times the correct value! These findings suggest that the hobbyist should exercise care in choosing test kits. Unfortunately, the most expensive kit was not the best, and neither was the cheapest kit the worst in terms of performance. If cost is not a satisfactory guide, then how is the inexperienced hobbyist to make his or her choice from among the many brands available?

The best approach is to make friends with an experienced aquarist whose success one can evaluate personally by observing his or her aquarium. If the tank looks clean and well-maintained, with crystal-clear water and fish swimming actively in vibrant health, chances are the novice should follow this person's recommendations regarding test kits. Joining the local aquarium club is probably the best way to meet experienced hobbyists. However, one might simply try starting a conversation with one of the other customers the next time he or she visits a favorite aquarium shop.

In the absence of a trusted fellow aquarist, ask the dealer. Good dealers will usually steer a hobbyist toward the best products on their shelves if the hobbyist lets them know that quality, and not merely the lowest possible price, is the goal.

For a given brand, results may not be equally satisfactory across the entire range of kits offered. The same company that makes a great nitrate test kit may have a lousy pH kit, for example. For this reason, you should avoid master test kit products that offer several kits in a single package. Here again, the best advice will come from experienced friends or a dealer.

When a novice decides to compare test kits, the usual approach is to analyze the same sample with different kits and compare the results. Superficially, this sounds like exactly what the author did in the evaluation described earlier. The key point, however, is that each kit must be compared not with the other kits but with an arbitrary standard, or control test, in order for the comparison to be valid. In the author's case, the control was the test performed by the professional lab based on the perfectly reasonable assumption that this would be the most accurate result.

Professional chemists evaluate test procedures by preparing a dummy sample, or *standard*, containing a precisely determined quantity of the substance of interest. The test is performed using this standard as the sample. The results are then examined to determine how closely they reflect the known value. One might wish that standard solutions for each of the substances of aquarium interest were on the market. Unfortunately, standard solutions must be freshly prepared before each round of testing. Simply allowing the solution to remain on the shelf for a period of time may alter its composition from the expected value due to evaporation, chemical changes, or other factors. For aquarists with access to the chemicals and equipment necessary, the use of standard solutions is the way to go. For the rest, however, there is little choice but to rely upon the recommendations of others. Since the test kits

TIP

Caution!

Never mix reagents from different tests or carry out any procedure not specifically recommended by the test kit manufacturer. Test reagents may be toxic, corrosive, or otherwise harmful. They must always be stored away from children and pets.

the author has used may not be available in all areas, there is little point in making specific brand recommendations.

Once a hobbyist settles on a kit for a particular measurement, he or she should stick with it. Changing brands will almost certainly mean that the numbers obtained will differ, making it impossible to compare results from the new kit with results previously determined with another kit. Since spotting a *changing trend* in tank conditions is more important than knowing the precise value at the current moment, the importance of having consistent test results cannot be overemphasized.

Testing Technique

Even the best test kit will not give good results if improperly used. The most common mistake amateurs make is to contaminate either the test vials or the chemical reagents, thus creating spurious test results. Following a few simple procedures will help ensure that the tests will be performed as accurately as possible.

Cleanliness is of utmost importance. Think about it. If the test is designed to determine

Above left: Tomato Anemonefish,
Amphiprion frenatus.

Above right: Blue Devil,
Chrysiptera cyanea.

Left: Golden Butterflyfish,
Chaetodon semilavartus.

Below: Rippled Coral Goby,
Gobiodon rivulatus.

*Always buy good-quality
test kits intended for marine
aquarium use.*

Leather Coral, Sarcophyton trocheliophorum.

Giant Clam, Tridacna maxima.

chemical concentrations in the range of parts per million, and most are, then small amounts of contamination can be disastrous.

Read the test kit label carefully for precautionary notes. Many reagents are toxic and should be handled with care. Also, never make the foolish mistake of using the top of the aquarium as a laboratory table. If a spill occurs, even a small amount of reagent entering the tank can have disastrous effects on the inhabitants.

To avoid confusion, do only one test at a time, record the results, and then proceed to the next test. Do not hurry. A complete set of tests for the typical marine tank should take only about half an hour to carry out.

Record Keeping

Do not spend the time to test the aquarium water without keeping a record of the results. Memory fails or plays tricks. Evaluating trends in water quality will be impossible without a complete record. The author keeps a looseleaf notebook handy for jotting down test results as they are performed. Later, this information can be manipulated with a personal computer to generate graphs that show how the tank is behaving as a function of time. These days, a hobbyist might use an HTML editor to record tank information, include photos captured with a digital camera, and post the data to a web page to share with other hobbyists. With the technological wizardry available to anyone with a computer, maintaining accurate and useful records should not present a problem.

Regardless of how it is done, keep track of the following: the date, time, and results of each test performed; the nature and amount of any water conditioner added; the amount of water change performed, if any; animal deaths or other circumstances that prompted the testing, if it is not routine. By maintaining such records over a period of time, the hobbyist will quickly learn to recognize the fluctuations in water parameters that occur normally versus those that suggest a problem. For example, as mentioned in the previous discussion regarding pH testing, the pH of the aquarium water varies depending upon the time of day because it is affected by photosynthesis, which occurs only in the light. Thus, at the end of the day, the pH is typically much higher than it is early in the morning. Such a diurnal fluctuation in pH is nothing to be concerned about. On the other hand, if the pH is measured at the same time of day on several occasions and is found to be declining, this is cause for action. Refer to page 53 for the details of this phenomenon. It clearly illustrates the importance of recognizing trends in water quality as opposed to performing spot checks randomly.

Instrumentation and Electronic Monitoring

Chemical test kits may give inaccurate results for a variety of reasons. For example, a color-blind aquarist may find it nearly impossible to distinguish subtle differences in the color of the indicator dye used to perform certain tests. Imprecise dispensing of reagents in the typical hobbyist kit can also introduce errors. To avoid such problems, certain parameters may be measured by means of electronic instruments. These are temperature, pH, salinity, dissolved oxygen, and redox potential. While these devices are a definite improvement over older methods of measurement, they must be properly calibrated to give accurate measurements. They must also be installed and maintained according to the manufacturer's recommendations in order to be considered reliable.

Handheld units for taking spot measurements offer the most economical alternative to chemical test kits and have the advantage of providing an instantaneous digital readout. Costlier units intended for continuous monitoring of aquarium conditions provide both a direct reading and a computer interface for extended data collection and manipulation. These multichannel devices can control lighting, pumps, and other equipment in addition to performing water analysis. At the time of this writing, handheld units for pH measurement cost about $100, while multichannel, programmable units cost about five times as much or more.

However, monitoring aquarium conditions electronically offers many advantages that may justify the added expense. For example, a valuable fish collection deserves the best temperature control system, as fluctuations can produce stress and precipitate an outbreak of disease. Stony corals should be maintained within a narrow range of pH. Electronic monitoring and control of pH results in greater stability than can be achieved by performing periodic tests and adding chemicals to adjust pH. Electronic units that can both monitor and control pH, temperature, and redox potential are particularly recommended for systems with greater than 100 gallons (400 L) of capacity and for those aquariums housing extensive collections of rare species.

Unfortunately, at present, there are no practical methods for electronic measurement of nitrogenous compounds, calcium, phosphate, or other dissolved substances that can be adapted to home aquarium use. Such specific ion determinations involve laboratory techniques and equipment that are beyond the capabilities of the typical home hobbyist. The probe cannot simply be immersed in the aquarium to take a reading as is the case with, for example, a pH meter.

Above left: Randall's Prawn Goby,
Amblyeleotris randalli.

Above right: Neon Dottyback,
Pseudochromis aldebarensis.

Left: Yellow Angelfish, Centropyge
flavissimus.

Below: Helfrich's Firefish,
Nemateleotris helfrichi.

Right: Auriga Butterflyfish,
Chaetodon auriga.

Middle left: Fairy Wrasse,
Cirrhilabrus cyanopleura.

*Middle right: Filament Flasher
Wrasse,* Paracheilinus
filamentosus.

Below: Picture Dragonet,
Synchiropus picturatus.

Regular water testing is really the only way to make certain that the chemical conditions of a marine aquarium are within acceptable parameters. Weekly testing is recommended. Little can be gained from the trouble of testing, however, if sloppy or improper techniques render the tests inaccurate. In order to be certain that the results obtained are reliable, the aquarist should keep in mind the simple procedures outlined below each time one checks the marine tank.

Make certain that all test equipment is properly cleaned. Before the first use, and after each subsequent test, thoroughly wash all glass or plastic utensils, test tubes, and so forth that may come into contact with the water sample. Use soap and warm water,

then thoroughly rinse, two or three times, in tap water. Follow this with a rinse in distilled water, since tap water may contain contaminants, too. Never wipe the test equipment. Instead, store it upside down on absorbent towels, allowing it to drain dry.

Before each test, rinse the vial with the water to be tested, then collect another sample upon which to perform the test.

Failure to follow the correct procedure for carrying out the test frequently leads to poor results. The aquarist must read carefully the instructions that came with the test, especially if he or she has not performed the test before or has switched brands of test kits. Not all nitrate tests, for example, are carried out the same way, although experi-

ence suggests that many hobbyists assume they are. Be sure to understand every aspect of the procedure thoroughly before beginning. Some tests, such as that for pH, for example, are quite simple. One simply adds a single reagent to the water sample and compares the resulting color with a chart that indicates the pH. In tests that measure ammonia, reagents are added. Then one must wait for a specified period of time before continuing with the next step. When encountering such a procedure, always use a timer, and allow for precisely the required waiting period.

Chemical tests are quantitative. The result always depends upon the degree to which the unknown component being measured reacts with the chemicals added during the test. It should be obvious, therefore, that one must add test reagents exactly as prescribed in the instructions. Too many hobbyists, however, make the mistake of approximating the amounts, especially after they have performed the same test a few times. Such shortcuts will inevitably produce an inaccurate reading.

Many tests require the aquarist to add reagents drop

Thoroughly clean all test equipment both before and after use.

by drop, either for a certain number of drops or until a color change occurs. When doing so, always hold the reagent bottle vertically. Squeeze with a gentle pressure to produce uniform drops in order to dispense the chemical as accurately as possible. The best-quality test kits use graduated pipettes and specify the number of milliliters of reagent to be mixed with the water sample. It should come as no surprise that such kits can be relied upon for more precise results than is possible with cruder, drop-count methods.

Tests that rely on counting drops until a color change occurs are titrations. In order to obtain the most accurate results possible, carry out the test in triplicate. Then average the three results obtained.

Other tests require one to compare the color obtained after the reagents have been added with a chart in order to determine the reading. Such tests are of two basic types. In some, such as pH tests, different readings produce different colors. Blue may mean a higher pH, while yellow may mean a lower one, and green may mean something in between, for example. As a rule, these tests are relatively easy to evaluate, unless one is visually impaired. More difficult to assess are tests in which the intensity of color varies with the concentration of the substance being measured in the sample. Typically, a pale tint means less of the substance, and a deeper tint means more is present. If the aquarist has trouble reading tests of this type, he or she should ask one or two other people to evaluate the result and compare their answers to his or her own. Tests for phosphate often fall into this category, largely because the amount one finds in the typical marine tank is quite low, and the resulting color in the test vial is faint.

Another reason for difficulty in reading the results of color-intensity tests is poor lighting.

Keep careful records of water test results, date, time, and any pertinent observations made.

Try to evaluate the test results in natural daylight from an east- or south-facing window. Fluorescent lighting may make greens and blues difficult to judge accurately, while incandescent lighting may do the same with reds and yellows. Dim light tends to obscure any color.

Always make note of test results in a log book. Record the date and time each test was performed. Commenting on any special circumstances that may have prompted the test is also helpful. As test results accumulate, plot the data onto graph paper. Graphing is especially helpful for pH and nitrate tests, as these parameters can be expected to change in a predictable fashion, although the exact pattern will vary from one aquarium to the next. After a couple of months, the graph will show the pattern typical of each tank. Any significant deviation can thereafter be taken as a sign of trouble that should be investigated.

APPENDICES

Tables on the following pages provide essential information regarding marine aquarium water chemistry. Refer to them when carrying out water tests.

Appendix A: Tables

Table 1: Marine Aquarium Water Quality Parameters

Marine environments are extremely stable, particularly where corals live. Significant deviation outside the range given for any of the parameters may lead to problems in the aquarium.

✔ Temperature—74°–82°F (23°–28°C)
✔ Salinity—34–36 ppt
✔ pH—8.15–8.6 (8.2–8.3 optimum)
✔ Alkalinity—2.0–5.0 meq/L (6–15 dKH)
✔ Ammonia (NH_3)—Zero
✔ Nitrite (NO_2^-)—Zero
✔ Nitrate (NO_3^-)—<20 mg/L (as nitrate ion)
✔ Nitrate (NO_3^-)—<4.55 mg/L (as nitrate-nitrogen)
✔ Phosphate (PO_4^{3-})—<0.05 mg/L
✔ Calcium (Ca^{2+})—375–475 mg/L
✔ Dissolved Oxygen (O_2)—>6.90 mg/L

Lionfish, **Pterois volitans.**

Six Spot Grouper, **Cephalopholis sexmaculata.**

Salinity Determination from Observed Specific Gravity Readings

Table 2: Conversion of Specific Gravity Readings to Density

Observed Hydrometer Reading	Temperature (°C)						
	20°	21°	22°	23°	24°	25°	26°
1.0170	10	12	15	17	20	22	25
1.0180	10	12	15	17	20	23	25
1.0190	10	12	15	18	20	23	26
1.0200	10	13	15	18	20	23	26
1.0210	10	13	15	18	21	23	26
1.0220	11	13	15	18	21	23	26
1.0230	11	13	16	18	21	24	26
1.0240	11	13	16	18	21	24	27
1.0250	11	13	16	18	21	24	27
1.0260	11	13	16	19	22	24	27
1.0270	11	14	16	19	22	24	27
1.0280	11	14	16	19	22	25	28
1.0290	11	14	16	19	22		

Table 3: Conversion of Density to Salinity

Density	Salinity	Density	Salinity	Density	Salinity
1.0180	25	1.0225	30	1.0270	36
1.0185	25	1.0230	31	1.0275	37
1.0190	26	1.0235	32	1.0280	38
1.0195	27	1.0240	32	1.0285	38
1.0200	27	1.0245	33	1.0290	39
1.0205	28	1.0250	34	1.0295	40
1.0210	29	1.0255	34	1.0300	40
1.0215	29	1.0260	35		
1.0220	30	1.0265	36		

Natural salinity is 35 in the vicinity of most coral reefs, unless the ocean is diluted from a nearby fresh-water source. The acceptable range for invertebrates is 34 to 36, while fishes can tolerate lower salinities.

Estimating Salinity from Observed Hydrometer Reading

Using a hydrometer, the salinity of the water can be calculated if the temperature is also known. First, convert the temperature to Celsius, by using the following formula:

$$°C = 0.56 \, (°F - 32)$$

Next, read across to find the temperature in Table 2, and then read down in that column to the row corresponding to the observed hydrometer reading. The number in the table is a conversion factor, which should be added to the observed hydrometer reading. The result is the density of the water.

Table 4: Selected Major and Minor Constituents of Natural Seawater

Element	ppm	Ionic Form
Chloride	19,500	Cl^-
Sodium	10,770	Na^+
Magnesium	1,290	Mg^+
Sulfur	905	SO_4^{2-}, $NaSO_4^-$
Calcium	412	Ca^{2+}
Potassium	380	K^+
Bromine	67	Br^-
Carbon[1]	28	HCO_3^{2-}, CO_3^{2-}, CO_2 gas
Nitrogen[2]	11.5	N_2 gas, NO_3^-, NH_4^+
Strontium	8	Sr^+
Oxygen	6	O_2 gas
Silicon	2	$Si(OH)_4$
Phosphorus[3]	0.06	PO_4^{3-}, HPO_4^{2-}, H_2PO_4
Iodine	0.06	IO_3^-, I^-
Molybdenum	0.01	MoO_4^{2-}
Iron	0.002	$Fe(OH)_2^+$, $Fe(OH)_4^-$
Copper	0.0001	$CuCO_3$, $CuOH^+$

[1] CO_2 gas is found in coral reef water at about 2 ppm. Dissolved organic carbon in surface seawater is seldom higher than 2.0 ppm and is generally much lower in coral reef habitats.

[2] Coral reef habitats typically have the lowest concentration of total inorganic nitrogen found in the ocean, in the range of only 1–2 ppb.

[3] In the water that bathes coral reefs, orthophosphate concentrations are frequently no higher than 0.1–0.2 ppb. The figure given is the average for the entire ocean.

Appendix B: Measurements and Equivalents

Measurements, Equivalents, and Formulas

Electricity

watts = volts × amps

amps = volts/watts

1 amp = 120 watts at 120 volts

30 amp line capacity = 3,600 watts (30 × 120)

20 amp line capacity = 2,400 watts (20 × 120)

15 amp line capacity = 1,800 watts (15 × 120)

Volume

1 gallon = 3.785 L (liter)

= 3,785 cc (cubic centimeter)

= 3,785 ml (milliliter)

1 liter = 1,000 ml

= 1,000 cc

= 0.264 gal

= 35.28 oz (weight)

= 33.8 fluid oz

= 2.25 lb

= 1 kg water

cc = ml

drop = 1/20 ml

ml = 20 drops

tsp = 5 ml

tbsp = 15 ml

fl. oz = 30 ml

Aquarium capacity (gal) = L × W × H (inches)/231

(liters) = L × W × H (cm)/1,000

(liters to gal) = × 0.264

Concentrations

parts per thousand (0/00 or ppt) = ml/L (liter)
                                         = 3.8 ml/gal
                                         = 3.8 gm/gal
                                         = gm/1,000 gms
                                         = gm/kilogram (Kg)
                                         = gm/2.2 lbs
part per million (ppm) = mg/L
                                         = 3.8 mg/gal
                                         = gm/cubic meter of water
                                         = ml/1,000 L
1 percent solution = 38 gm/gal
                                         = 38 ml/gal
                                         = 1 oz/3 quarts
                                         = 1.3 oz/gal
                                         = 10 gm/L
                                         = 1 gm/100 ml
                                         = 10 ml/L
1:1,000 solution = 1 ml/L
                                         = 0.1 gm/100 ml
                                         = 3.8 gm/gal
                                         = 3.8 ml/gal
                                         = 0.13 oz/gal

Temperature

Centigrade or Celsius $= (F - 32) \times 0.55$
Fahrenheit $= (C \times 1.8) + 32$

Dissolved Gases

Oxygen in ppm $\times 0.7 = O_2$ in cc/L or ml/L
Oxygen in ml/L or cc/L $\times 1.429 = O_2$ in ppm

CO_2 in ppm $\times 0.509 = CO_2$ in ml/L or cc/L
CO_2 in cc/L or ml/L $\times 1.964 = CO_2$ in ppm

Logarithm to the Base 10 Prefix Terms

billion	10^9	(G)	giga	1,000,000,000
million	10^6	(M)	mega	1,000,000
thousand	10^3	(k)	kilo	1,000
hundred	10^2	(h)	hecto	100
ten	10	(dk)	deka	10
(tenth)	10^{-1}	(d)	deci	0.1
(hundredth)	10^{-2}	(c)	centi	0.01
(thousandth)	10^{-3}	(m)	milli	0.001
(millionth)	10^{-6}	(μ)	micro	0.000001
(billionth)	10^{-9}	(n)	nano	0.000000001
(trillionth)	10^{-12}	(p)	pico	0.000000000001

Alkalinity and Water Hardness

Several synonyms for "alkalinity" are used in aquarium literature. Alkalinity is expressed in milliequivalents per liter (meq/L). Other terms used are "carbonate hardness" expressed in parts per million (ppm) or grains per gallon (gr/gal) of calcium carbonate, and "German hardness," or "KH," expressed in degrees (dKH). The relationships among these units are:

$$1 \text{ meq/L} = 50 \text{ ppm } CaCO_3 = 2.92 \text{ gr/gal } CaCO_3 = 2.8 \text{ dKH}$$

Descriptive terms for water hardness found in aquarium literature are subject to varying interpretation. We have indicated approximate ranges of hardness corresponding to each term.

very soft	=	<4 dKH or <75 ppm $CaCO_3$
soft	=	5–8 dKH or 80–150 ppm $CaCO_3$
medium hard	=	9–12 dKH or 160–220 ppm $CaCO_3$
hard	=	13–20 dKH or 230–360 ppm $CaCO_3$
very hard	=	>20 dKH or >360 ppm $CaCO_3$

Light Intensity

1 lumen	=	1 footcandle
1 lux	=	1 lumen/square meter

Length

1 micron	=	0.001 millimeter
1 millimeter	=	0.001 meter = 0.039 inches
1 centimeter	=	0.01 meter = 0.39 inches
1 meter	=	39.37 inches = 3.28 feet
1 foot	=	30.48 centimeters = 0.3 meters
1 yard	=	3 feet = 0.91 meters

Surface Area

1 square centimeter	=	0.16 square inches
1 square meter	=	10.76 square feet = 1.2 square yards
1 square inch	=	6.45 square centimeters
1 square foot	=	144 square inches = 0.09 square meters
1 square yard	=	9 square feet = 0.81 square meters

Weight

1 milligram	=	0.001 grams
1 gram	=	1,000 milligrams = 0.035 ounces
1 kilogram	=	1,000 grams = 35 ounces = 2.2 pounds
1 ounce	=	28.35 grams
1 pound	=	16 ounces = 454 grams = 0.45 kilograms

Other Data

1 gallon of fresh water weighs 8 pounds

1 gallon of seawater weighs 8.5 pounds

1 cubic foot of water = 7.5 gallons = 37.8 liters

GLOSSARY

Listed on the following pages are the terms appearing in italics in the text. Formal definitions are included to facilitate understanding of the material presented in this book.

Accuracy: the degree to which a test reflects the actual condition of the sample being tested

Acid: a chemical compound that donates one or more hydrogen ions in a chemical reaction

Aerobic: requiring oxygen to sustain life

Alkalinity: a measurement of the resistance of a solution to a change in pH as acid is added

Anaerobic: capable of living in the absence of oxygen or incapable of surviving in the presence of this gas

Anion: a negatively charged atom or molecule

Atom: the smallest unit of a chemical element that retains all the properties of that element

Atomic weight: the sum of the weights of the protons and neutrons in an atomic nucleus

Base: a chemical compound that accepts one or more hydrogen ions in a chemical reaction

Biochemical cycling: an ecological process in which the waste products of one organism serve as a food source for another organism

Biological filtration: the detoxification of nitrogenous wastes in aquarium water, mediated by nitrifying bacteria

Bloom: sudden, uncontrolled growth, as of algae

Oceanic Seahorse, **Hippocampus kuda.**

Buffer: a solution that is resistant to changes in pH despite small additions of acid or base

Calcium hardness: a measure of the amount of calcium dissolved in a water sample

Calibration buffer: a solution of precisely known pH used for standardization of pH meters and test kits

Calorie: the amount of heat required to raise the temperature of 1 gram of water by 1 degree Celsius

Carbonate hardness: a synonym for alkalinity, expressed in parts per million of calcium carbonate equivalents

Carrying capacity: a measure of the total amount of living organisms that can be maintained by a life support system, such as that of an aquarium

Cation: a positively charged atom or molecule

Chemical bond: the force that holds two or more atoms together, forming a compound

Chemical reaction: the interaction between two or more atoms or molecules, resulting in the production of new combinations unlike the original reactants

Complex ion: one composed of two or more different atoms

Compound: a substance formed when two or more elements undergo a chemical reaction

Conserved ions: the 11 major ions found in natural seawater, the ratios among which do not vary as the salinity changes

Covalent bond: a chemical bond in which two atoms share a pair of electrons between them

Denitrification: the process of conversion of nitrate to nitrogen gas, mediated by anaerobic bacteria

Denitrification potential: the overall capacity of a closed system to convert nitrate into nitrogen gas

DOC: dissolved organic carbon

Electronegativity: the tendency for an atom or molecule to gain electrons

Element: a substance composed of only one type of atom

Equilibrium: said of any system in which two opposing processes exactly balance each other, resulting in a steady state

Equivalent: the amount of acid or base contained in a normal solution; 1 equivalent of any acid will neutralize 1 equivalent of a base

Food web: a community of photosynthetic and nonphotosynthetic organisms that are ecologically related to each other as prey and predators

Hardness: a measure of the dissolved mineral content of water

Hydrogen bond: a weak interaction among water molecules in which the two positive ends of each are attracted to the negative ends of its neighbors

Hyperosmotic: having body fluids more salty than the surrounding water

Hypertonic: a synonym for hyperosmotic

Hypoosmotic: having body fluids less salty than the surrounding water

Hypotonic: a synonym for hypoosmotic

Ion: an atom or molecule bearing a net positive or negative charge as the result of having gained or lost one or more electrons

Ion-exchange resin: a product capable of removing dissolved ions from water, usually replacing them with sodium ions

Ionic bond: a chemical bond in which one or more electrons are exchanged between two atoms

Live rock: rock taken from the ocean and sold for aquarium use with its natural complement of living organisms more or less intact

Live sand: sand taken from the ocean floor and sold for aquarium use with its natural complement of living organisms more or less intact

Mole: a gram-molecular weight of any element or compound; 1 mole of any substance contains 6.02×10^{23} particles

Neutralization reaction: one in which an acid and a base combine to form water

Nitrification: the two-step process by which ammonia in aquarium water is converted to nitrate, mediated by nitrifying bacteria

Nitrogen cycling: the exchange of compounds containing nitrogen among organisms in a food web

Nitrogen equilibrium: said of an aquarium in which the amount of nitrate detectable in the water remains constant with time

Normal solution: one containing 1 equivalent of an acid or base

Orthophosphate: the chemical form of phosphate, abbreviated PO_4^{3-}, that is measured by water test kits

Osmoconformer: an organism with body fluids that reflect the saltiness of the surrounding water

Osmosis: the movement of water molecules across a semipermeable membrane, such as that of a living cell

Osmotic pressure: the tendency for water to move across a semipermeable membrane from an area of higher salt concentration to an area of lower concentration

Oxidation: the loss of electrons by an atom or molecule

Oxidizing agent: the electron acceptor in a redox reaction

pH: the degree of acidity or alkalinity of a solution, equal to the negative logarithm of the hydrogen ion concentration

Photosynthesis: the process whereby green plants use light energy to form food molecules from carbon dioxide and water

Polar: said of a molecule having both positive and negative ends

Precision: referring to the smallest increment detectable by a test; the smaller the increment, the greater the precision

Protein skimmer: an aquarium filtration device used for removal of dissolved organic carbon by injection of fine air bubbles into a column of water; organic matter is sequestered in the foam thus produced and is trapped in a reservoir from which it can be periodically emptied

Reagent: any substance participating in a chemical reaction

Redox reaction: one in which 1 participating atom or molecule gains electrons as another atom or molecule loses them

Reducing agent: the donor of electrons in a redox reaction

Reduction: the gain of electrons

Salinity: the total amount of dissolved substances in seawater

Specific gravity: the ratio of the weight of a given volume of any liquid to the weight of an equivalent volume of pure water

Specific heat: the number of calories necessary to raise the temperature of 1 gram of any substance by 1 degree Celsius

Standard: a solution of known concentration against which the accuracy of a chemical test may be evaluated

Strong acid: one in which the dissociation of the molecule into its component ions is essentially complete under normal conditions

Supersaturation: containing a higher proportion of a dissolved gas than is theoretically possible

Titration: a chemical test carried out by adding a reagent of known concentration in small increments to a sample until the sample has completely reacted with the reagent

Trophic level: the position of a particular organism in a food web

Weak acid: one in which the dissociation of the molecule into its component ions is not complete under normal circumstances

Books

Adey, Walter H., and Karen Loveland. *Dynamic Aquaria*. New York: Academic Press, 1991.

Delbeek, J. Charles, and Julian Sprung. *The Reef Aquarium*, Volume One. Coconut Grove, Florida: Ricordea Publishing, 1994.

Fossa, S., and A. J. Neilsen. *The Modern Coral Reef Aquarium*, Volume One. Bornheim, Germany: Birgitt Schmettkamp Verlag, 1996.

Goldstein, Robert J. *Marine Reef Aquarium Handbook*. Hauppauge, New York: Barron's Educational Series, Inc., 1997.

Moe, Martin A., Jr. *The Marine Aquarium Reference. Systems and Invertebrates*. Plantation, Florida: Green Turtle Publications, 1989.

Spotte, Stephen. *Captive Seawater Fishes*. New York: John Wiley & Sons, 1992.

Sprung, Julian, and J. Charles Delbeek. *The Reef Aquarium*, Volume Two. Coconut Grove, Florida: Ricordea Publishing, 1994.

Tullock, John. *Natural Reef Aquariums*, Shelburne, Vermont: TFH/Microcosm, 1999.

Anemone Shrimp, **Periclimenes brevirostris.**

Internet Sites

For subscribers to America Online: Keyword AQUARIUM provides links to major public aquarium web sites. Go to Keyword HOBBIES, then select "Aquarium Science" from the menu to access the aquarium hobby message board.

For subscribers to Compuserve Information Service: GO FISHNET accesses the huge forum for aquarists, both freshwater and marine, including a library of indispensable information on a host of topics.

Other sites to visit using any web browser:
Breeder's Registry:
http://www.actwin.com/fish/br/index.html
(A web site for propagators of marine fish and invertebrates.)

Center for Marine Conservation:
http://www.cmc-ocean.org

Coral Reef Alliance: *http://www.coral.org*

Coral Reef Task Force:
http://coralreef.gov/trade.html

Ecovitality: *http://www.ecovitality.org*

Geothermal Aquaculture Research Foundation:
http://www.garf.org

Harbor Branch Oceanographic Institute:
http://www.hboi.org

Marine Aquarium Council:
http://www.aquariumcouncil.org

Striped Puffer, Arothron manilensis.

Online Reef Aquarist Society:
  http://www.panix.com/~cmo1/reef/

Reeflink: *http://www.reeflink.com*

Newsgroups
news.alt.aquaria
news.rec.aquaria
news.sci.aquaria

Periodicals
SeaScope
Aquarium Systems, Inc.
8141 Tyler Boulevard
Mentor, OH 44060

Journal of Maquaculture
Breeder's Registry
P.O. Box 255373
Sacramento, CA 95865

Aquarium Fish Magazine
Fancy Publications
P.O. Box 6050
Mission Viejo, CA 92690

Coral Reef, Reef Encounter
International Society for Reef Studies
Dr. John Ogden
Florida Institute of Oceanography
830 First Street S
St. Petersburg, FL 33701

Ocean Realm
4067 Broadway
San Antonio, TX 78209

Freshwater and Marine Aquarium Magazine
RCM Publications
144 West Sierra Madre Boulevard
Sierra Madre, CA 91024

Tropical Fish Hobbyist
TFH Publications, Inc.
One TFH Plaza
Neptune, NJ 07753

Safety Around the Aquarium
  Water and electricity can lead to dangerous accidents. Therefore, you should make absolutely sure when buying equipment that it is also really suitable for use in an aquarium.
✔ Every technical device must have the UL sticker on it. These letters give the assurance that the safety of the equipment has been carefully checked by experts and "with ordinary use" (as the experts say) nothing dangerous can happen.
✔ Always unplug any electrical equipment before you ever do any cleaning in or around the aquarium.
✔ Never do your own repairs on the aquarium or the equipment if there is something wrong with it. As a matter of principle, all repairs should only be carried out by an expert.

Cover Photos

Aaron Norman

Photo Credits

Aaron Norman: pages 2-3, 4, 5, 8 (top left), 9 (top left and right), 12 (bottom and right), 13 (left, top, and bottom), 16 (top left, top right, and bottom left), 17 (top left), 20 (top, bottom left, and bottom right), 24, 25, 28 (bottom left and bottom right), 29 (bottom left and right), 32 (top right and bottom), 33 (top right and bottom left), 36 (top right), 37 (top left and right), 40, 41, 44 (top left, top right, and bottom left), 45, 48 (top and bottom), 49 (bottom left), 56, 57, 60 (top right and middle left), 61 (top right, bottom left, and bottom right), 64 (bottom left), 65 (top left, middle left), 68, 69, 72 (top left, top right, middle, and bottom), 73 (bottom left and bottom right), 76 (top left and right), 80, 81 (top and bottom), 88, 89, 92, and 93; David Schleser: pages 28 (top left) and 44 (bottom right); Mark Smith: pages 8 (top right and bottom), 9 (bottom left and right), 12 (top left), 16 (bottom right), 17 (top right, middle left and right, and bottom), 20 (middle right), 21 (top left and right, middle left and right, and bottom), 28 (top right), 29 (top left and right), 32 (top left), 33 (top left), 36 (top left, middle left, middle right, and bottom), 37 (bottom), 49 (top left, top right, and bottom right), 52, 53, 60 (top left, middle right, and bottom), 61 (top left and middle right), 64 (bottom right), 65 (middle right and bottom), 76 (middle and bottom), and 77 (top, middle left, middle right, and bottom).

Important Note

Electrical equipment for aquarium care is described in this book. Please do not fail to read the note on page 93, since otherwise serious accidents could occur.

Water damage from broken glass, water over-flows, or tank leaks cannot always be avoided. Therefore, you should not fail to take out insurance.

Please take special care that neither adults, children, nor pets ever eat any aquarium plants. Doing so can cause substantial health risks. Fish medication should be kept away from children and pets.

About the Author

John Tullock is a writer, photographer, and avid aquarist. He earned a master's degree in biology and has been keeping aquariums for over 35 years. He has owned a retail and mail-order aquarium business and is the founder of the American Marinelife Dealers Association. He served on the steering committee that created the Marine Aquarium Council and has addressed the Global Biodiversity Forum on aquarium trade issues. He is the author of seven previous books and scores of articles for aquarium hobbyists.

All inquiries should be addressed to:
Barron's Educational Series, Inc.
250 Wireless Boulevard
Hauppauge, NY 11788
http://www.barronseduc.com

International Standard Book No. 0-7641-2038-7

Library of Congress Catalog Card No. 2001043299

Library of Congress Cataloging-in-Publication Data
Tullock, John H., 1951–
   Water chemistry for the marine aquarium /
   John H. Tullock.
        p. cm.
   Includes bibliographical references (p.).
   ISBN 0-7641-2038-7 (alk. paper)
   1. Marine aquariums. 2. Water chemistry. I. Title.
SF457.1 .T865   2002
639.34′2—dc21   2001043299

Printed in China

9 8 7 6 5 4 3 2